George E. Parris

Thermodynamics:

Heat Capacity,

Enthalpy, Entropy

&

Free Energy

Lectures by

George E. Parris

3rd Ed. April 2019

Preface to the 3rd Edition

As noted in the 1st edition (2015) and 2nd Edition (2017) this lecture is targeted at molecular scientists, not engineers or physicists. I have done my best to demystify the concepts described here. In my opinion, textbook authors since 1900 have rarely understood Clausius, Maxwell, Boltzmann, Gibbs or Nernst and have resorted to hand-waving arguments, which obscure understanding. I think it is useful to examine the history of these concepts because it cuts through some of the mystery.

In the 2nd edition, I have generally cleaned up the original text, added references to clarify and support the original discussion; summarized the historical development of enthalpy and entropy, and included supplemental material on the statistical interpretation of entropy. I also mention the methods by which enthalpy and entropy are normally determined experimentally. In the 3rd edition I have extended the vibrational concept to heat capacity to simple salts (which are analyzed as moles of ions) and use this to show examples where the heat capacity is essentially constant (0-298°K) and thus the heat capacity at 298°K is the average heat capacity (0-298°K) and hence is the standard entropy (S_{298}). This is one of the principal conclusions of Clausius.

A surprising development came out of my analysis (in the 2nd edition) when I considered the case where the temperature of a system approaches absolute zero. I think the result is applicable to the phenomenon of superconductivity. However, a superficial reading of some of the theory of solid-state physics leaves me uncomfortable with the results. The solid-state physicists use an entirely different vocabulary and it is not clear to me at this time where these theories merge with chemistry (as they must). The phenomenon of superconductivity appears to be a phase change involving only the electrons. When the motions of the ion-cores that make up metals and semiconductors drop below a certain level (e.g., vibrational energy in the ground state at least transverse to the direction of current), the electrons have undisturbed continuous paths of travel; hence, the de Broglie wavelengths (i.e., wave functions) of the electrons approach the physical dimension of the conductor and the conductor becomes superconducting.

Criticisms of the manuscript are welcomed.

George E. Parris

Gaithersburg, MD

March 30, 2019

1.0 Basic Terminology and Concepts

When physical changes or chemical reactions occur, one system is converted into a different system. These processes are accompanied by changes in the energy distribution within the system and (usually) between the system and surroundings as shown in the following diagram.

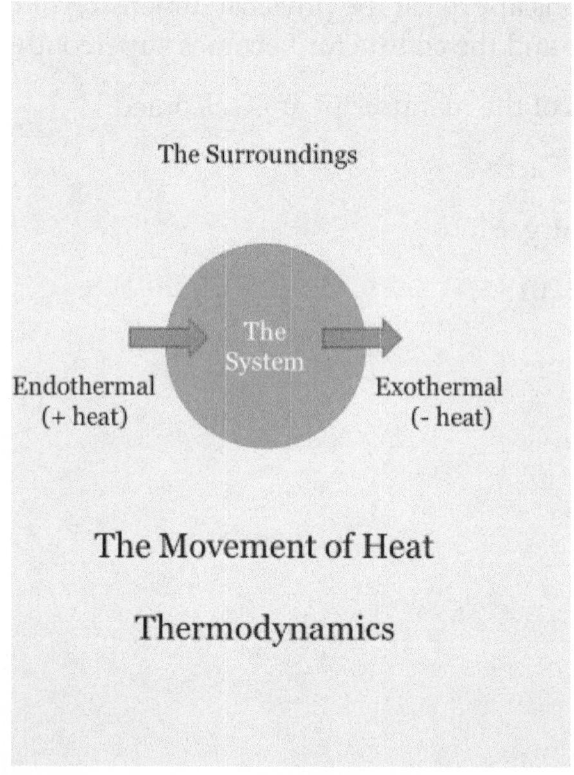

There are some terms and concepts that we need to agree on before we start:

1.1 Heat

Heat is kinetic energy, i.e., energy resulting from movement of molecules and atoms (see 2.0 below). Recall for an *ideal monatomic gas*[1]:

Heat/n =

Kinetic Energy per mole = (3/2)RT

Note that ***heat energy is directly proportional to absolute temperature.***

Heat **spontaneously flows** from areas of high energy density (i.e., high temperature) to areas of low energy density (i.e., low temperature) by conduction, convection or radiation.

1.2 System and Surroundings

The **system** is whatever collection of atoms and molecules we are talking about (we may define it any way we wish); the **surroundings** is everything else in the universe (see figure above).

When heat enters the system, we say the movement is "**endothermal**" and assign the movement of heat a positive (+) sign.

When heat leaves the system, we say the movement is "**exothermal**" and assign it a negative (-) sign.

---

[1] If you are not familiar with the kinetic theory of gases, I suggest you begin by reading my book on that topic also published on Amazon/kindle.

If a process occurs at a constant temperature (either because it did not involve heat or was maintained at a constant temperature by exchange of heat with the surroundings) we say it was "**isothermal.**"

When there is no (net) exchange of heat between the system and the surroundings in a process (e.g., a chemical reaction), we say the process is "**adiabatic**." We can make a process adiabatic by insulating the system from the surroundings.

An Adiabatic Container

(Dewar's Thermos Flask) Photograph by LepoRello, source Wikimedia Commons

An adiabatic container is approximated by having a vacuum between the system and the surroundings (to minimize conduction and convection) and using a silvered mirror surface (to reflect radiation). Note that, processes that happen very fast (e.g., combustion or an explosion) approximate an adiabatic process (even if the temperature difference between the system and the surroundings is large) because the movement of heat energy may be slow relative to the chemical and physical change.

1.3 Heat Capacity

Heat capacity is the ability of a molecule to store heat at a specific temperature (T).

Heat Capacity (T) = Δ(Heat Contained /n) /ΔT

= ΔKinetic Energy *per mole*/ΔT

Specific heat is a similar term, but it is usually measured in Heat Contained **per gram** of material:

Specific Heat (T) = ΔHeat Contained *per gram*/ ΔT

Note that heat capacity and specific heat are functions of temperature.

Also note that energy can be stored by phase changes (which ideally occur at constant temperature). Phase changes contribute to the energy stored in a system and thus to the average heat capacity of a system over a defined temperature range (e.g., 0°K to T°).[2]

2.0 Molecular Dynamics

In the analysis of ideal gases, *monatomic* gases (point masses) were assumed. Gases like helium, neon, argon, and mercury

---

[2] **T°** is used to refer to an *arbitrarily defined* standard temperature (e.g., 298°K).

vapor closely approximate this model because they are monatomic and have little atomic affinity. But, most gases (e.g., water, carbon dioxide, methane) are not monatomic. The basic gas laws still apply (e.g., PV = nRT); but the heat capacity of these gases is different because of the ways they can store kinetic energy.

Consider a *monatomic gas;* the only way it can store energy is by moving its *center of mass* from point to point (i.e., **translation**):

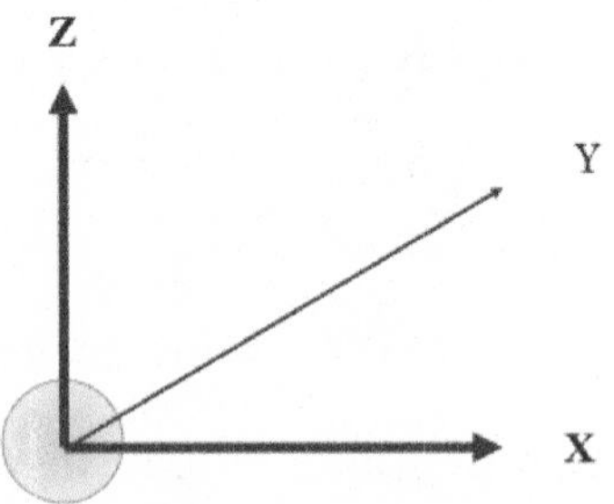

Every particle has three independent **"degrees of freedom"**

That is, it can move in any of three independent directions in 3-dimensional space.

Thus, we say that a monatomic molecule **has three translational degrees of freedom**. Notice that when a particle translates, it moves its **center of gravity**.

Now consider, a diatomic molecule like O_2, N_2, Cl_2 or CO.

Consider a **diatomic molecule**:

Two particles held together with a chemical bond

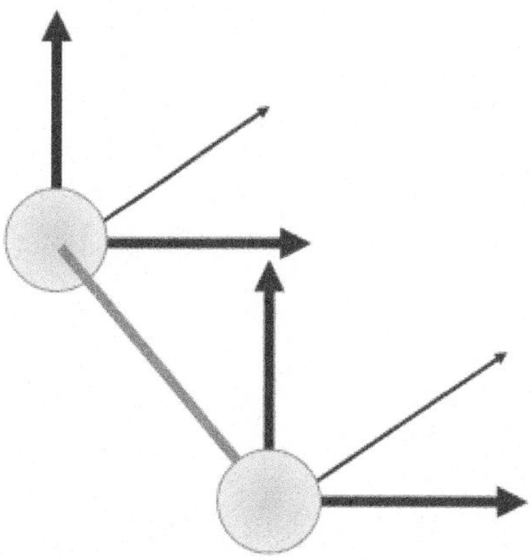

Remember, very particle has three independent
"degrees of freedom"

So the molecule must have
2 x 3 = 6 degrees of freedom.

What are these degrees of molecular freedom? Well, three of the degrees of freedom are **translations** that move the center of gravity of the molecule *without changing the bond distance or orientation of the molecule.*

But the molecule can also rotate about two axes *without moving the center of gravity *CG)or changing the bond lenght:*

A diatomic molecule can **rotate**
Around two axes
Without changing
(i) the position of the
center of gravity (CG) Or
(ii) the bond length:

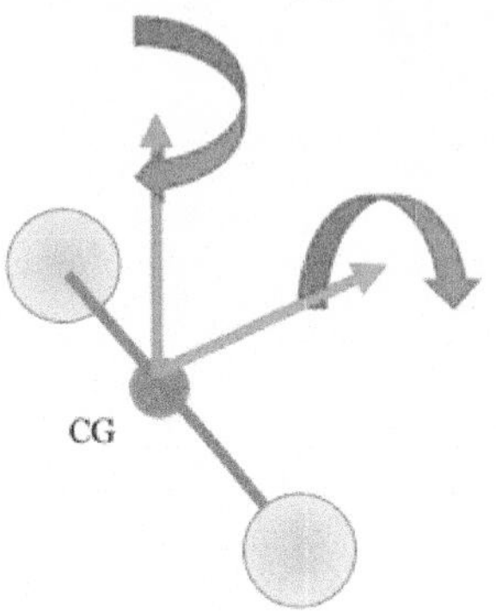

These two rotations
(around the center of gravity)represent
two molecular degrees of freedom.

These rotations are characterized by the *moment of inertia* of the molecule.

Let us keep score:

2 x 3 = 6 total degrees of freedom

- 3 translations

- 2 rotations

1 degree of freedom

There is still 1 degree of freedom not accounted for; it does not move the center of gravity and it is not a rotation: It can only be a vibration:

A diatomic molecule can vibrate without changing its center of gravity:

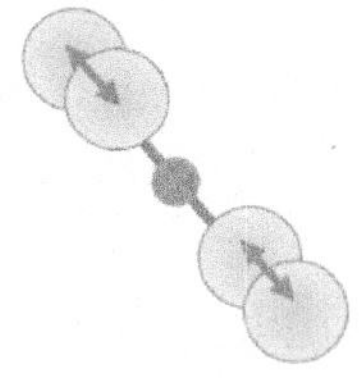

The vibration *(change in bond length)* represents one molecular degree of freedom.

Thus, for a diatomic molecule (note that all diatomic molecules are linear):

2 x 3 = 6 total degrees of freedom

- 3 translations

- 2 rotations

1 vibration

Note that *all diatomic molecules are linear* and will only have two rotational degrees of freedom. Some triatomic molecules (e.g., CO_2) are also linear and even one tetratomic molecule (HCCH, acetylene) is linear.

But most polyatomic molecules are *non-linear and have 3 degrees of rotational freedom.*

Thus, for a non-linear triatomic molecule like water (H_2O)

3 x 3 = 9 total degrees of freedom in H_2O

- 3 translations

- 3 rotations

3 vibrations

(i.e., asymmetric stretch, symmetric stretch and bend)

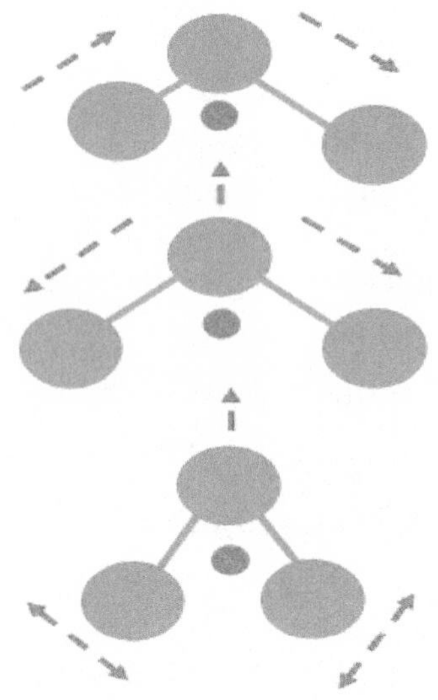

3.0 Quantum Mechanics, Spectroscopy and Heat Capacity

3.1 Photons and Quantized Energy Levels

On the scale of atoms and molecules, the energy that can be stored in any mechanical system is not continuous, but rather it is restricted to certain energy levels that are separated by discrete differences in energy; i.e., the *systems are quantized.* Similarly, electromagnetic radiation (e.g., visible light) carries energy in discrete packages called *photons* where the energy of the photon (Ep) is equal to the product of Planck's constant (h) and the frequency (ν). Note that the frequency is equal to the speed of light (c) divided by the wavelength (λ). Thus, the energy of a photon is directly proportional to frequency (measured in cycles/s) and directly proportional to the inverse of the wavelength (1/λ), which is called "wave number" (typically measured in the units of cm^{-1}).

Recall that energy must be conserved instantaneously. Thus, only photons of energy (hν) that correspond *exactly to the energy level separation* of some physical system can be absorbed. Also note that interconversion of electromagnetic (photons) and mechanical (kinetic) energy requires that there are changes in the electric or magnetic polarization of the mechanical system corresponding to the electric and magnetic vectors of the photon.

Obviously, with the absorption of a photon, the molecular system must move to the higher energy state. The lowest available energy state is called the ground state (G_0) and the other energy levels are called excited states (G_i where i = 1, 2, 3, …).

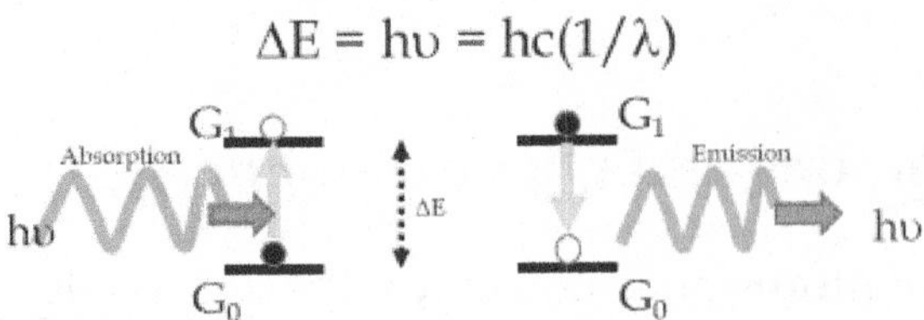

Similarly, if a molecular system is in an excited energy state, it may fall to a lower state and in this process the system generates an electromagnetic impulse (i.e., photon) that carries away the energy as electromagnetic radiation (as shown above).

Each degree of freedom of a molecule, represents a mechanical system quantized in a specific pattern of energy levels. The electronic structure of atoms and molecules also involves quantized energy levels. The reasons for the patterns (e.g., energy level separations) is beyond the scope of the current discussion. However, we can make general statements:

(1) Electronic energy levels of atoms and molecules are typically separated by energy on the order of ~10^6 J/mole for tightly bound electrons to ~10^3 J/mole for loosely held electrons (e.g., rearrangement of pi-electrons or ligand field splitting of d-type orbitals). Most single-bond energies are in the range of 10^5 to 10^6 J/mole (10^2 to 10^3 kJ/mole). Thus, some single-bonds (sigma-molecular orbitals) may be broken by photons in this range (ultraviolet light). Pi-molecular orbitals are separated by ~10^4 to ~10^5 J/mole and frequently absorb visible light.

(2) Vibrations of molecules typically have energy level separations ~10^3 J/mole.

(3) Rotations of molecules typically have energy level separations ~10^2 J/mole.

(4) Translations of molecules typically have energy level separations of ~10^1 J/mole.

Mode of Energy Storage	Temperature Needed to Significantly Populate the First Excited Level	Typical Energy Level Separations
Electronic	>3000°K	10^{-18} J/molecule *Ultraviolet Radiation*
Vibration	300°K	10^{-20} J/molecule 500 to 5000 cm^{-1} *Infrared Radiation*
Rotation	30°K	10^{-21} J/molecule 50 to 500 cm^{-1} *Microwave Radiation*
Translation	3°K	10^{-23} J/molecule

These energies can be compared to thermal energy (RT) at the corresponding absolute temperatures (R = 8.314 J/mole °K).

Notice that since most (large) molecules decompose below 3000°K, electronic excitations generally do not play a role in thermodynamics. However, a few very stable small molecules (CO_2, CO) and some free radicals (parts of molecules) and all individual atoms are stable at 3000 °K and we can expect to see electronic excitations in these molecules and atoms.

For example, if you put potassium ions into a flame, you see a characteristic color and pattern of line spectra associated with the atomic electronic excitations:

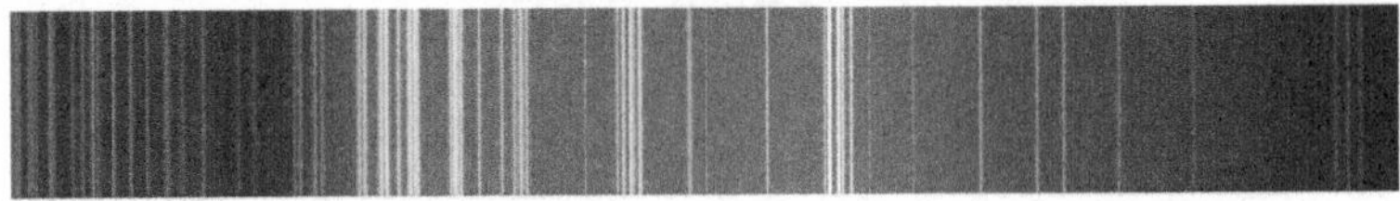

https://upload.wikimedia.org/wikipedia/commons/d/d0/Potassium_Spectrum.jpg

3.2 Spectroscopy

Spectroscopy is the technique of studying the structure and motions of molecules by using photons to probe the energy-level structure of the molecules. In practice, there are many types of spectroscopy. One of the most useful involves using infrared light (usually measured in frequency as wave numbers, 4000 to 400 cm^{-1}) to study the vibrations of molecules. Here, IR spectroscopy is presented as a relevant example because a substantial amount of heat energy is stored in the vibrations of polyatomic molecules.

Consider formaldehyde (H_2CO) in the gas phase:

4 x 3 = 12 total degrees of freedom in H_2CO

- 3 translations

- 3 rotations

6 vibrations

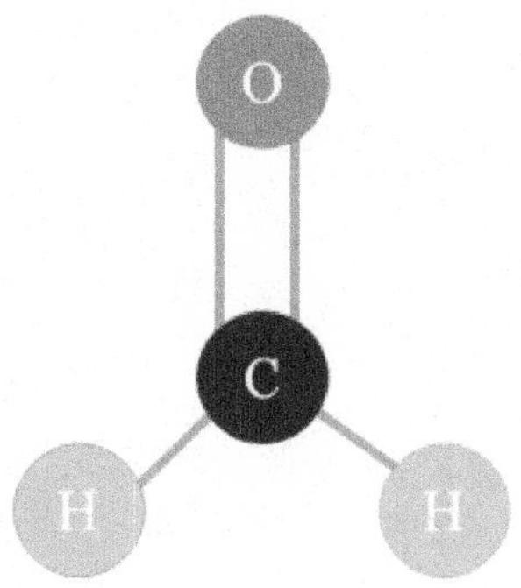

The vibrations can be observed by absorption of infrared radiation because, in each vibration, the dipole moment changes and can interact with the electric vector of the photon.

What you see in the IR spectrum are the 6 vibrational transitions:

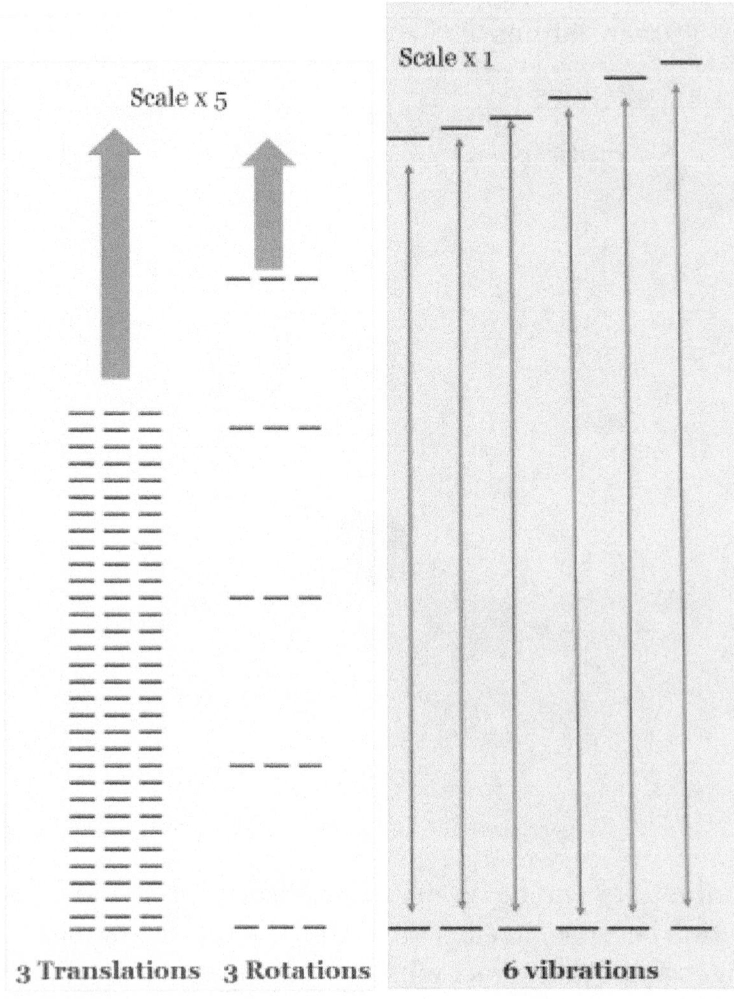

The resulting spectrum is shown below:

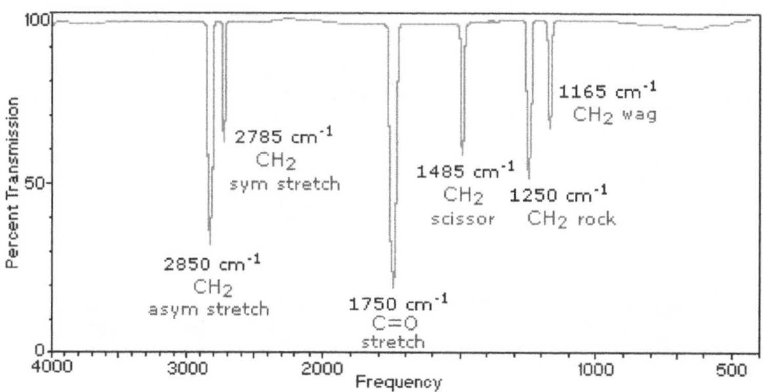

Source:
https://www2.chemistry.msu.edu/faculty/reusch/VirtTxtJml/Spectrpy/InfraRed/infrared.htm

Note that at ambient temperature (~300°K) most molecule are in vibrational ground states and what is observed is excitation to the first excited level of each degree of freedom. Keep in mind that at ambient temperature, these molecules (in the gas phase) are more or less randomly distributed among a very large number of translational and rotational energy levels (magnified in the diagram by a factor of 5).

It has been stated above that in order for an exchange of energy between a photon and a physical system to occur (i) the photon energy and the energy difference of the energy states must match (exactly) and (ii) there must be a mechanism for the electric or magnetic vector of the electromagnetic radiation to interact with the molecule (e.g., a change in dipole moment when the change in energy levels occurs). Carbon dioxide (a linear molecule) provides an interesting example:

Carbon Dioxide is linear

O=C=O 3 x 3 = 9

-3 translations

<u>-2 rotations</u>

4 vibrations

(A, B, C, D)

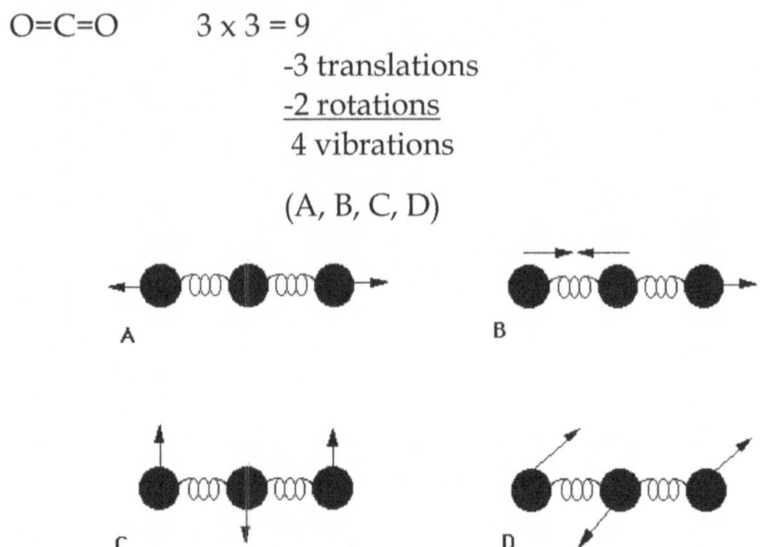

http://www.wag.caltech.edu/home/jang/genchem/infrared.htm

The C=O bonds are polar, but notice that overall the molecule is non-polar (the polarities O=C and C=O cancel out). Thus, it is also observed that the symmetric stretching vibration (A) does not change the dipole moment of the molecule and hence will not absorb electromagnetic radiation as shown in the IR spectrum of CO_2 below:

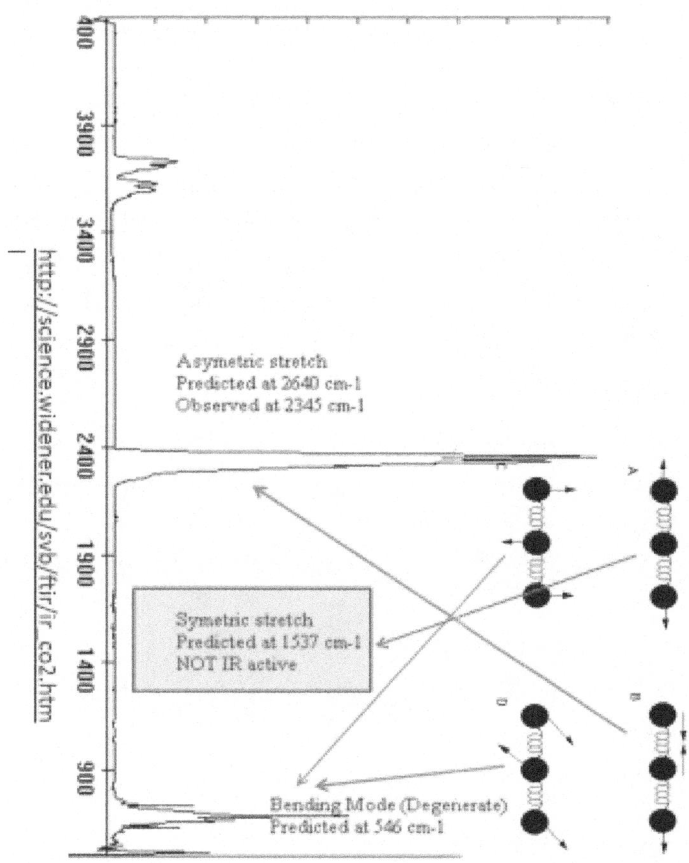

The red arrow in the figure above shows where the symmetrical stretch would be expected (based on energy level separations) but it does not absorb radiation there.

3.3 Boltzmann's Distribution

Ludwig Boltzmann (1844-1906) was a statistician who applied techniques developed by James Clerk Maxwell

(1831-1879) to physical phenomena. In particular, he addressed the problem of how kinetic energy would be distributed among the various degrees of freedom (each with its own energy separation pattern) of a collection of molecules. Assuming random distribution (Gauss's law) perturbed (i.e., skewed) by an exponential function, Boltzmann and Maxwell produced a statistical description of mechanics.

If we consider a single degree of freedom with evenly spaced energy levels (ε_i), the ratio of the number of molecules with energy in the ith energy level to the number of molecules in the ground state (ε_0) is given by the equation:

$$N_i/N_0 = g_i/g_0\ e^{-\Delta\varepsilon/RT}$$

Where

-Δε is in J/mole,

R is the universal gas constant (8.314 J/mole °K)

g_i is the "degeneracy" of the ith energy level (i.e., when there are two or more energy levels with the same energy, they are said to be *degenerate*).

Based on the discussion in section 3.2 above, we would expect that a number (upwards of 10) rotational energy levels would be occupied to different degrees (N_i/N_0) at ambient temperature (~300°K). Through an interesting coincidence, we can in fact observe them.

In a polyatomic molecule, the vibrations and the rotations do not affect each other very much because they are independent (i.e., they are not tied together…i.e., changing the average length of a bond has little affect the angular momentum around any axis). In contrast, in a diatomic molecule, changing the average length of a bond has a large and direct effect on the angular momentum around the center of gravity:

The Moment of Inertia must change as the bond-length changes

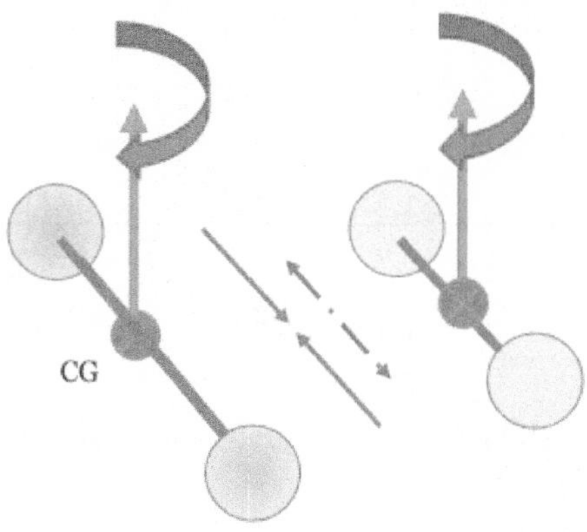

The vibrations and rotations must be **coupled** (i.e., must happen at the same time) for diatomic molecules.

Thus, in a diatomic molecule, the *vibrations and rotations are coupled* because you cannot change the average bond length without changing the average moment of inertia. When you observe the vibrational spectrum of HCl in the *gas phase,* you find a pattern like this:

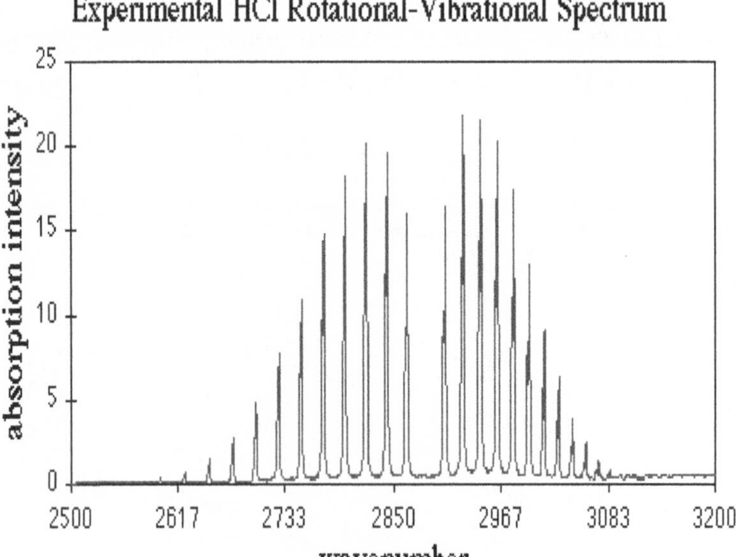

http://faculty.evansville.edu/bl22/images/programming/exphcl.gif

The resolution of this particular spectrum is not high enough to completely resolve the stretching vibrations of the isotopes of chlorine ($H^{35}Cl$ and $H^{37}Cl$), which give the ragged appearance to the peaks. Nonetheless, what we are looking at here is the vibrational transition (energy level $\mathbf{V_0}$ to $\mathbf{V_1}$ nominally at about 2885 cm^{-1} in the the IR range) *coupled* with changes in the rotational energy levels (+/- Δ**J**) where the rotational energy level **J** is transitioning *simultaneously* with the vibrational transition and thus changing the net $\Delta E = \Delta E_v +/- \Delta E_j$ as reflected in the observed wave number ($1/\lambda$).

In the context of this discussion, the most interesting point here is that the intensities of the transitions show the relative occupancy (N_i/N_0) of the rotational energy levels, which decrease exponentially as the rotational energy increases as predicted by the Boltzmann equation (above). Note that the pattern of intensities is temperature dependent.

3.4 Storage of Kinetic Energy in Vibrations

At ambient temperature (~300°K), the translational and rotational energy levels are not randomly occupied, but enough of them are occupied that little additional heat storage capacity is obtained at higher temperature. In contrast, at ambient temperature, most molecules are just beginning to utilize vibrational energy levels for storage of kinetic energy. Thus, as a practical matter, we need to consider how the energy stored in vibrations increases as the temperature increases. This is accomplished by a quantum mechanical calculation. Typically, the calculation (that is not done here) examines the single vibration of a diatomic molecule. This result is equivalent to assuming that each vibration of a complex molecule is independent of all the others (i.e., the vibrations do not couple). The result is shown as:

$$\textbf{Vibrational Energy} = \mathbf{RT\,[x/(e^x - 1)]}$$

where $\mathbf{x = \varepsilon/RT}$ (ε is in J/mole)

Obviously, for a complex molecule with many vibrational degrees of freedom, you would have to do this calculation for each degree of freedom, which is characterized by its own (spectroscopically determined) energy level separation (ε). Tables and graphs of the vibrational energy for various values of the energy level spacing (ε) and temperature (T) have been calculated and can be found in reference books.

For our purposes here, you should realize that bending vibrations usually have smaller energy level separations than stretching vibrations and thus begin to absorb significant energy between 300°K and 1000°K (200 cm^{-1} to 700 cm^{-1})), while the stretching vibrations absorb significant energy between 500°K and 4000°K (350 cm^{-1} to 3500 cm^{-1}). The following graph gives an idea of the effect of temperature on heat capacity for a vibrational degree of freedom:

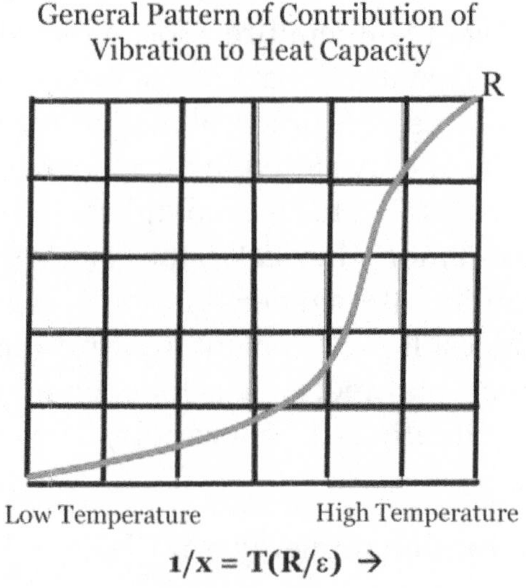

3.5 Heat Capacity of Real Compounds

Chemical compounds store kinetic energy in their mechanical degrees of freedom and their electronic transitions[3].

Recall from the discussion of ideal (monatomic) gases that heat capacity at constant volume (Cv) is

$$\mathbf{Cv = (3/2)\ R}$$

And that the heat capacity at constant pressure (Cp) is equal to the Cv plus the work done to expand the medium:

$$\begin{aligned} \mathbf{Cp} &= \mathbf{Cv + P\Delta V/nT} \\ &= \mathbf{Cv + R} \\ &= \mathbf{(3/2)R + R} \\ &= \mathbf{(5/2)R} \end{aligned}$$

Recall that R = 8.314 J/mole °K

Thus, **Cp for an *ideal monatomic gas* = 20.785 J/mole °K**

This calculation is very close to the experimentally observed value for a wide range of monatomic gases including: the inert gases (which vary in mass for 4 AMU to 222 AMU), hydrogen atoms (1 AMU), sodium atoms, etc.

---

[3] Electronic transitions are actually potential energy but they correspond to extreme extensions of chemical bonds ("hyper-vibrations") and thus energy flows freely from vibrations to electronic transitions.

Look at the heat capacity of carbon dioxide over a range of temperatures (Table below). The total heat capacity (Cp) is the sum or all the different degrees of freedom (Cv_i) plus the work (PV) term:

Cp =

Cv(trans) + Cv(rotat) + Cv(vib) + Cv(elect) + R

Keep in mind that general rule that below 30°K rotations, vibrations and electronic degrees of freedom are not available for energy storage (i.e., some say that they are "frozen out"). Below 300°K, vibrations and electronic degrees of freedom are frozen out; and below 3000°K electronic energy levels do not contribute appreciably to energy storage. Thus, while it is difficult to identify specific degrees of rotational and vibrational freedom[4], the data can be analyzed in terms of Cv(translation), Cv(rotation) and Cv(vibration) and Cv(electronic).

For quantum mechanical reasons (beyond the scope of this topic) the maximum contribution of various degrees of freedom is as follows:

	Maximum contribution to Cv	**Temperature Range**	
		Minimum Contribution	Maximum Contribution

	per degree of freedom	°K	°K
Translatios	**½ R**	**3**	**≥100**
Rotations	**½ R**	**30**	**≥200**
Vibrations	**R**	**300**	**≥2000**

[4] In the case of CO_2 (a linear molecule) the two rotational degrees of freedom are degenerate, and we can assume that energy flows into the less energetic vibrations first.

Experimental heat capacity (Cp) of carbon dioxide

Temperature °K	Cp Joule/mole°K	Source of Cp
175	31.2	**Translations and Rotations** small contribution from bending vibrations
200	32.3	
300	37.2	**Bending and stretching vibrations** contribute to heat capacity
400	41.3	
500	44.6	
600	47.3	
700	49.5	

800	51.4	
900	52.9	
1000	54.3	
2000	60.3	vibrations saturated
2500	61.4	
3000	62.2	
4000	63.2	some electronic excitation
6000	64.9	

Overall, the heat capacity (Cv) *always* increases as the temperature increases, because more energy levels become available for storage of energy. Between 3 and 2000°K the mechanical systems provide the energy storage systems; above 2000°K electronic transitions also contribute. The PV work term is constant (R) at all temperatures at constant pressure.

In liquids, the heat capacity includes the enthalpy of breaking intermolecular bonds. In particular, intermolecular hydrogen bonds vary widely (e.g., ~2 kJ/mole to ~160 kJ/mole). The hydrogen bonds in water (~21 kJ/mole) break down by 400°K and account for a large element in the heat capacity and heat of vaporization of water.

3.6 Absolute Zero

Before leaving this topic, it should be noticed that by defining the energy levels of all degrees of freedom, we can provide a clear definition of *absolute zero* temperature:

> *Absolute zero is the state in which a molecule is in the lowest available energy level in all degrees of freedom.*

This does NOT mean that all molecular motion stops at absolute zero.

You may have heard of the Warner Heisenberg's (1901-1976) quantum mechanical principle of "uncertainty." Basically, you cannot simultaneously know the position and momentum of a particle with absolute accuracy. If all vibration of a molecule stopped, you *could* know the position and momentum with absolute accuracy. Thus, if you look at a typical potential energy curve for the stretching of a chemical bond, you will find that the *lowest available energy level* still allows some vibration to occur. Thus, the classical definition of absolute zero should be modified as follows:

Absolute zero it the temperature when all ***available*** *kinetic energy has been removed from a molecule.*

The origins of IR energy levels in molecules

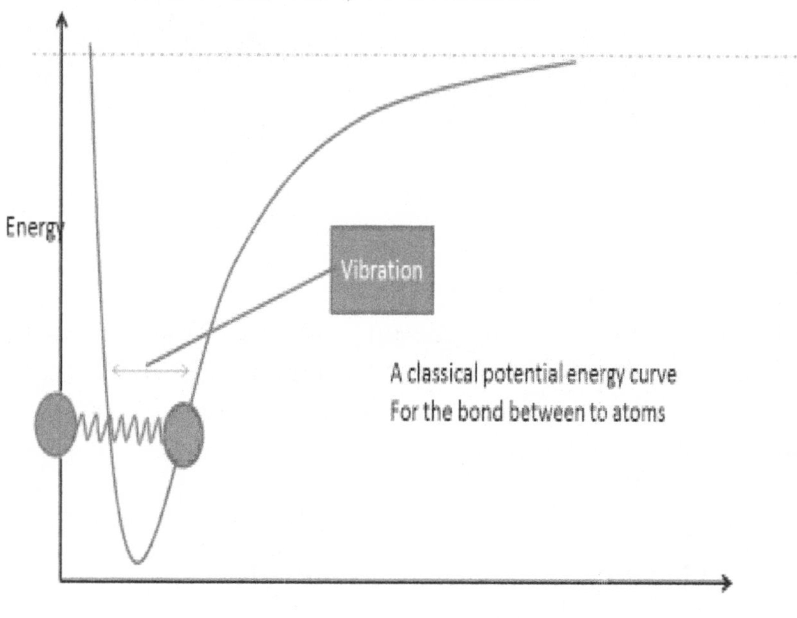

r, Interatomic distance

The "zero point" vibrational energy remains in the molecule.

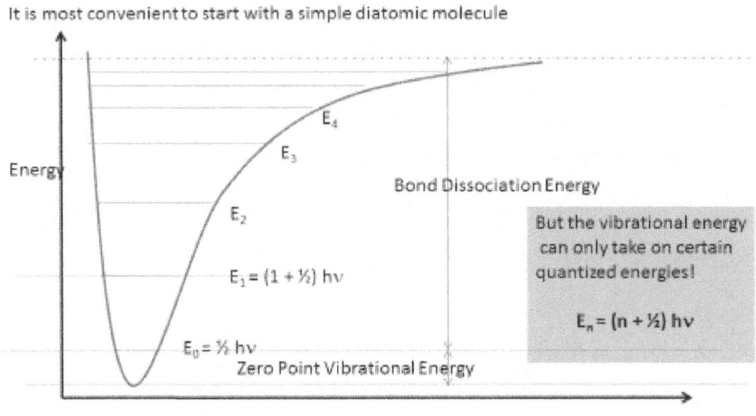

Notice that as the bond nears the breaking point, the vibrational energy level become more closely spaced and eventually merge with translational energy levels. If the potential energy function were a true parabola, the energy level spacing would remain constant.

3.7 The Uncertainty Principle

The Heisenberg uncertainty principle is an important concept in quantum mechanics. It is generally stated that we cannot simultaneously know the position and momentum of a small particle (e.g., atom or electron). Ironically, Warner Heisenberg originally formulated the idea based on our inability of measure certain values. But, his mentor Niels Bohr (1895-1962) took exception to the idea of basing a fundamental principle on limitations of our monitoring equipment and added substantial footnotes to the proof of the paper that was submitted by Heisenberg (March 1927). Although the concept is taught in virtually every modern chemistry textbook, no relevant examples are provided.

The "zero point" vibrational energy is an excellent demonstration and proof of this principle. If the vibrational ground state was at the minimum of the potential energy curve, then we could predict the position and momentum of the atoms at absolute zero. But, since the vibrations still continue (in a narrow range of bond lengths) when the molecule is in the vibrational ground state, we cannot assign a precise quantity to its position and momentum simultaneously.

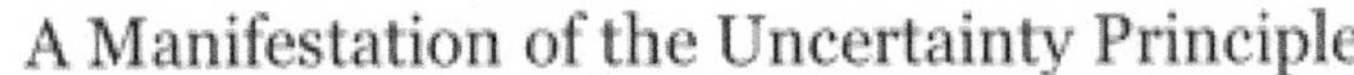

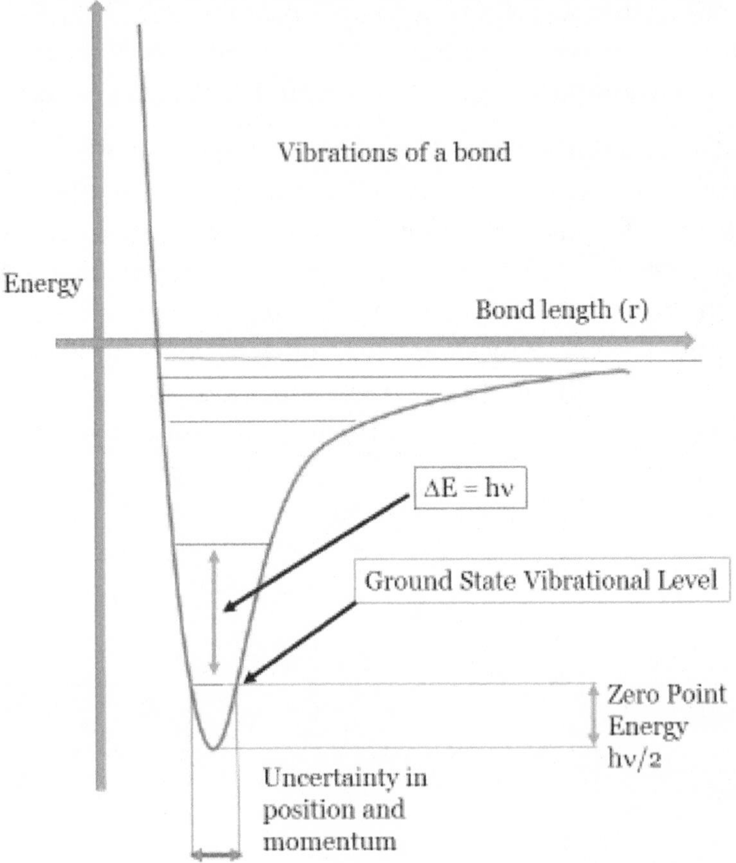

3.8 The Heat Capacity of Metals and Simple Salts

For the most part we have ignored the liquid or solid phase in the discussion above. Obviously, in the solid phase, atoms and molecules cannot translate, and rotation may be restricted (unless the molecules is highly symmetrical. Nonetheless, molecules can "bounce" around a central position. This movement is called *libration* and can take the form of partial rotations (in polyatomic molecules) or vibrations in monatomic materials like metals. As a result, the heat capacities of all metals are very near 3 R (i.e., 24.9 J/mole°K). For example, consider these metals:

Metal	Specific Heat (J/g°K)	Atomic Mass	Heat Capacity (J/mole°K)
Li	3.56	6.941	24.71
Na	1.23	22.99	28.28
K	0.75	39.10	29.33
Fe	0.444	55.85	24.80
Cu	0.385	63.55	24.47
Ag	0.240	107.9	25.90
Au	0.129	197.0	25.41

The excess heat capacity may be attributed to movement of the electrons, which are the principle carriers of electric current and thermal conductance.

A simple salt (e.g., NaF, density = 2.56 g/mL) is more organized than a metal (e.g., Na, density = 0.97 g/mL). Although the fluoride ions are less massive (19 amu) than the sodium ions (23 amu), NaF is more than twice as dense than sodium metal. The volume of the metal occupied by free electrons is now occupied by fluoride ions and they the array of ions is more compact. Nonetheless, the salt still fits the model of a metal if we consider the motion for *individual particles* (ions). Thus, we would predict a heat capacity of simple salts to be 3 R *per mole of particles*. In the case of NaF, one mole of the salt has two moles of particles (ions); thus, on a molar basis of the salt, the heat capacity of NaF should be 2 x (3 R) ≈ 50 J/mole °K. Indeed, the experimental value is 46.82 J/mol °K.

Other simple salts are similar:

Salt	N x 3R (J/mole °K)	Cv at 298°K (J/mole °K)
NaF	49.9	46.8
NaCl	49.9	50.5
NaBr	49.9	51.4
NaI	49.9	52.1
MgF_2	74.8	61.6
$MgCl_2$	74.8	71.1
$MgBr_2$	74.8	70.0
AlF_3	99.7	75.1
$AlCl_3$	99.7	91.1
Li_2O	74.8	54.1
Na_2O	74.8	73.0
MgO	49.9	37.2
Al_2O_3	124.7	89.7

But, as the crystal structured become more complicated, the different sites in the lattice have different force constants and store energy differently. Strongly bonded groups of atoms may only vibrate as a group at moderated temperature (i.e., there are fewer vibrational degrees of freedom available at this temperature (298°K) than predicted by simply counting the atoms). This particularly apparent in the oxides and fluorides, which have very high lattice energies and low heat capacities at ambient temperature (see above)

4.0 The Absolute Energy of a System is Always Conserved

Energy may be stored in various forms (kinetic, potential, photonic, etc.), but it is always conserved instantaneously. I can think of two obvious confirmations that energy is conserved instantaneously: (1) Spectroscopy is based on the assumption that only photons (hυ) that match the energy difference of two physical states *exactly* can be absorbed. Photons with more or less energy pass through the system. (2) Blackbody radiation is the result of collision of physical particles. In this case momentum must be conserved; but if you do the math for a few simple systems, you will realize that when you conserve momentum, the particles involved have less kinetic energy after the collision. Thus, a photon is released that carries away the excess kinetic energy.

4.1 The Absolute Energy of a System

The Absolute Energy of a system is the sum of all the energy that goes into a collection of molecules assuming that the starting point (i.e., zero energy) is a collection of individual atoms in space surrounded by a medium at constant pressure (P). Absolute energy at a given standard temperature (T^o) is composed of three parts:

(i) the potential energy of bond formation ($\mathbf{E_0}$) at absolute zero

(ii) all the thermal (kinetic) energy associated with raising the molecules from absolute zero (T = 0°K) to some standard temperature (T°). This includes thermal energy and energy associated with phase changes. It must be recognized that the heat capacity of real systems is not constant, it is a function of temperature, Cv(T).The thermal energy can be calculated by multiplying the average heat capacity in the range 0 to T° (i.e., Cv_{avg}(0-T°)) by T°.
To find the *average heat capacity over the temperature range (0-T°)* the method of Clausius (see below) is used. Multiplying by the standard temperature (T°) gives the absolute heat content at T°.

$$\textbf{Thermal Energy} = T^{o} \times \int_{0}^{T_o} (Cv(T)/T^{o})\, dT$$

This definition *necessarily includes* the thermal energy (i.e., kinetic energy) currently stored in the molecular system (translation, rotation, vibration) at the standard temperature (T°).

(iii) the work required to create a space (i.e., volume, V) for the molecule in the medium (at constant pressure = P).

$$\textbf{work} = P(\Delta V)$$

In the figure below, I have summarized the contributions to absolute energy in a graphical format. In the graphic, I define the "zero" energy state as the separate (point mass) atoms with no kinetic energy (except that required by the uncertainty principle). The atoms are thus at absolute zero (T

= 0 and V = 0) under a *constant* external pressure of (P). The first step in this process is to form molecules from the atoms. Formation of bonds (from isolated atoms) is always and exothermal process and the potential energy released is known as the bond energy at absolute zero (E_0). To maintain the system at absolute zero, all of this energy must be released from the system to the surroundings (as shown by the broad arrow).

Then we put thermal energy into the molecules. This requires an input of energy, which is calculated by multiplying their Cv x ΔT (also shown by a broad arrow). If we assume that the molecules are ideal gases, then Cv is a constant and note that ΔT = T- 0 = T. *But, for real substances the Cv is a function of T; and in this figure, the **average Cv** of the temperature range (T = 0 to T = T^o) as defined by Clausius (see below).*

Finally, the system of molecules at ambient temperature is allowed to expand against the external pressure to a non-zero volume (ΔV = V – 0 = V). This requires and input of work energy (PΔV = work) (also indicated by a broad arrow).

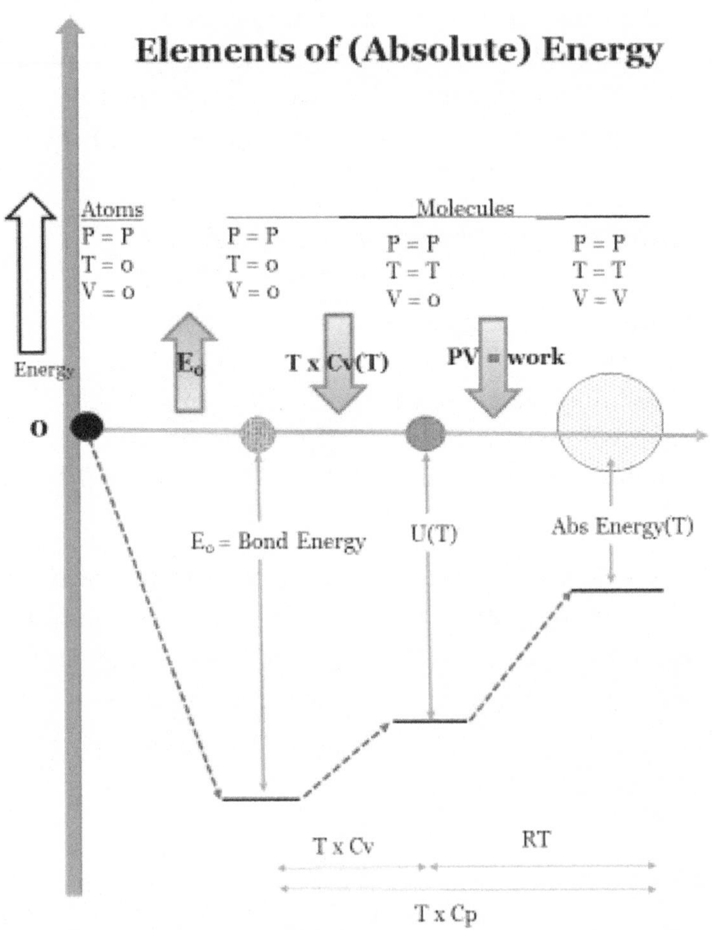

In principle, summation of these three terms represent all the relevant energy (i.e., heat energy)[5] in the system at ambient temperature.

---

[5] There is energy that is not accounted for such as nuclear binding

When we say that energy must be conserved, this is the energy we are talking about. The *absolute energy* determines the position of chemical equilibria.

4.2 Clausius and Entropy

4.2.1 The Molecular Theory of Gases and Kinetic Theory of Heat

In the early 1800s, the engineers (James Watt (1736-1819), Sadi Carnot (1796-1832) and James Joule (1818-1889)) were dominating the public's attention because of their practical applications of energy and heat. Indeed, the nature of gases was generally ignored except for the *empirical relations* developed by Robert Boyle (1627-1691), Jacques Charles (1746-1823) and Louis Gay-Lussac (1778-1850).

$$\mathbf{R = PV/nT}$$

The *caloric theory* of heat (which replaced the phlogiston theory in the early 1800s) treated heat as a *material fluid* that flowed from place to place and was widely embraced by the engineers. This theory required no molecular explanation. The mathematical model of gases that was based on molecular kinetic energy (1738) of Daniel Bernoulli (1700–1782) was little appreciated outside of Russia and the independent efforts of the scientific outsider John Herapath (1790–1868) in 1820 were totally ignored.

In 1843, John James Waterston (1811–1883) proposed a similar kinetic model of gases and his timing was better than Herapath. Up until that point (1850) the caloric theory[6] of

---

energy etc. that are outside the realm of chemistry and chemical reactions.

heat was unchallenged. Waterston's paper attracted the attention of Hermann von Helmholtz (1821-1894) who passed it to August Karl Krönig (1822–1879) who published a widely recognized kinetic theory of gases in 1856. With a kinetic molecular interpretation of gases, Bernoulli's interpretation of the relationship of temperature and energy were soon to follow.

Meanwhile, Julius Robert Mayer (1814–1878) is credited with articulating the concept of conservation of energy in 1842 (J. R. von Mayer, *Annalen der Chemie und Pharmacie* 43, 233 (1842)) and he calculate the mechanical equivalent of heat before James Joule (1843). We can see a parallel of discoveries in chemistry/physics and engineering during this time. The engineers generally had higher visibility because their applications led directly to valuable practical applications. Trained in physiology, Helmholtz came to similar conclusions as his contemporaries (e.g., Hess, Mayer and Joule) regarding conservation of energy and published a paper discussing conservation of energy in muscle contraction in 1847. These advances in understanding of energy, work and heat essentially defined the *first law of thermodynamics* and set the stage for quantitative applications.

4.2.2 Absolute Zero and Entropy

Rudolf Clausius (1822-1888) was among the young scientists who were trying to explain the phenomena of energy that was literally driving the industrial revolution in the early

---

[6]The caloric theory (which replaced the phlogiston theory) treated heat as a material fluid that flows from place to place, and this is adequate for most engineering purposes even today.

1800s. With the idea of conservation of energy (i.e., the first law of thermodynamics) taking hold in the 1840s, Clausius (1850) published a paper entitled "On the Moving Force of Heat and the Laws of Heat, which May Be Deduced Therefrom." He was interested in phase changes and discussed "free heat" and "latent heat." Clausius articulated a corollary regarding the movement of heat spontaneously from hot to cold (the second law of thermodynamics) in 1854-56.[7] Krönig's 1856 paper on gases attracted the attention of Clausius, who was beginning to see heat as kinetic energy of molecules. Clausius (1857) expanded the kinetic model by recognizing that multi-atom gases have rotational and vibrational kinetic energy as well as translational kinetic energy and introduced the concept of "mean free path."

The idea of an absolute limit to temperature had been raised periodically since Robert Boyle (1627–1691); and in 1779, Johann Lambert (1728-1777) gave a reasonably accurate value of -270°C. But, on the basis of a variety of arguments from imminent chemists and physicists, far different numbers (or no limits at all) were proposed. (Without equating heat to kinetic energy, why should there be a limit to how much could be extracted from a material?) It was not until 1848 that Lord Kelvin [William Thomson, 1st Baron Kelvin] (1824–1907) explicitly embraced the idea that heat was kinetic energy and convinced the academic community that there was, indeed, an absolute zero of temperature (i.e., no available energy) and that absolute zero was -273°C. This

---

[7] Clausius, R. 1856. On a Modified Form of the Second Fundamental Theorem in the Mechanical Theory of Heat. *Philos. Mag*. 4. 12 (77): 81–98.

event created a marker from which thermal energy could be measured *absolutely* and undoubtedly influenced Clausius.

It is clear that Clausius was turning these issues of absolute temperature, phase changes, heat capacity and conservation of energy over in his brain and undoubtedly came to the conclusion that when we consider a system at ambient temperature, we must account for *all the energy* that went into it to raise it from absolute zero (0°K). But, it was not until 1865 that Clausius suggested a general mathematical form for the thermal energy involved in taking a system from absolute zero (T = 0°K) to its current state (at ambient conditions, T°). He may have been motivated to call this factor to everyone's attention because of the contemporary interest being shown to *experimentally determined enthalpy* of reactions (as noted below).

Recognizing that energy (in the form of heat) is added to a system as its temperature is raised and recognizing that real systems go through a variety of phases and phase changes between absolute zero and ambient temperature, Clausius opted to generalized the calculation by treating heat capacity as a variable function of temperature and integrating heat capacity Cv(T) = (dq/dT) over the temperature range of interest (0 to 298°K):

$$\mathbf{Cv_{average}(0\ to\ T^{o}) = \int_{0}^{To} (Cv(T)/T^{o})\ dT}$$

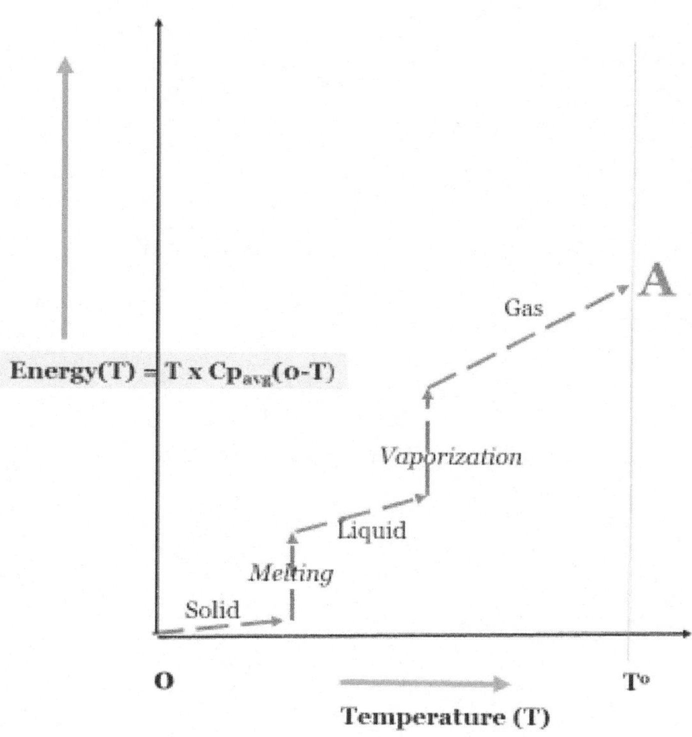

In the graph above, it is shown that a typical material may go through two (major) phase changes (at constant temperatures T_{fp} and T_{bp}) before it reaches ambient temperature (T^o) and increases it temperature between the phase changes. Note that the energy incorporated in the system at any temperature (T) is determined by the *average*

heat capacity over the range (0-T). Thus, the energy incorporated in the system at (T°) can be calculated:

$$\text{Energy}(T^{o}) = T^{o} \times \int_{0}^{T^{o}} (Cp(T)/T)\, dT$$

$$\text{Energy}(T^{o}) = T^{o} \times S(T^{o})$$

Clausius called the integral "entropy" (S) and he[8] goes on to say:

> *"If we wish to designate* **S** *by a proper name we can say of it that it is the* <u>*transformation content*</u> *of the body, in the*

[8] *Ueber verschiedene fur die Anwendung bequeme Formen der Hauptgleichungen der mechanischen Warmetheorie published in Annalen der Physik und Chemie, 125, 353- (1865) (as translated by William Frqancis Magie)*

same way that we say of the quantity **U** *that it is the heat and work content of the body. However, since I think it is better to take the names of such quantities as these, which are important for science, from the ancient languages, so that they can be introduced without change into all the modern languages, I propose to name the magnitude* **S** *the* <u>*entropy*</u> *of the body, from the Greek word ητροπη, a transformation. I have intentionally formed the word* <u>*entropy*</u> *so as to be as similar as possible to the word energy, since both these quantities, which are to be known by these names, are so nearly related to each other in their physical significance that a certain similarity in their names seems to me advantageous."*

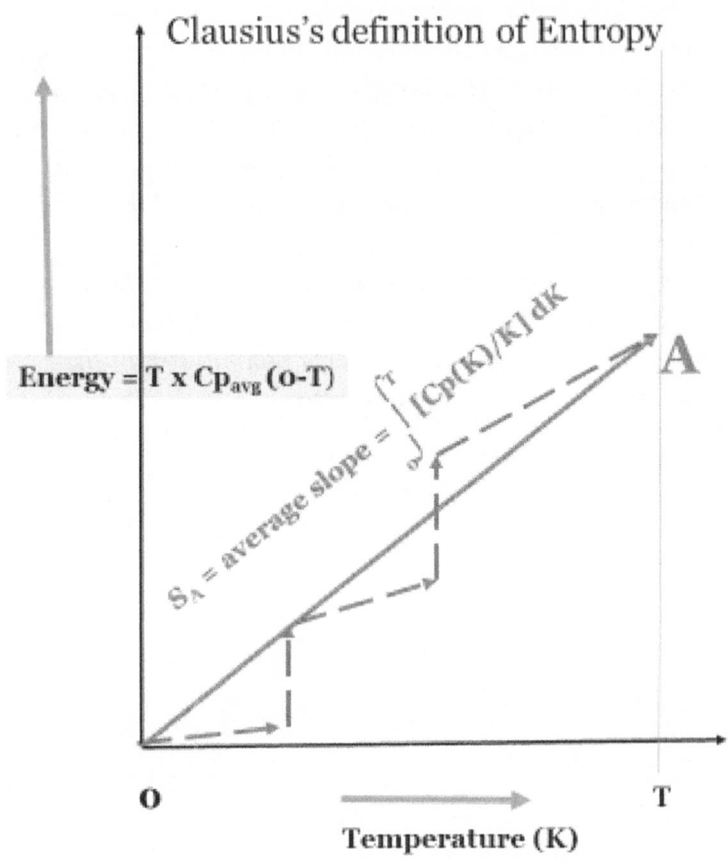

Entropy (S) is the average heat capacity from o to T.
© Parris 2015

It should be obvious that the term T x (S(0-T)) as defined by Clausius[9] is the absolute amount of thermal energy (plus PV)

[9] Clausius called the intrinsic molecular characteristic S(0-T) the "transformation factor." But, I believe he should have called it

that I have called out in the absolute energy defined in section 4.1 (above).

The equivalency of the average heat capacity (Cv avg) and entropy (S) may not be obvious to some readers. However, if consider a crystal of a simple salt (e.g. NaF, NaCl, NaBr, Na_2O, $MgCl_2$, MgB_{r2}) we would expect the heat capacity to change very little on average between 0°K and 298°K (i.e., no phase change). Thus, Cv should be constant over that range and thus the Cv at 298°K is virtually identical to the Cv avg over the range. As a result, the Cv at 298°K is very similar to the molar entropy (S_{298}) at T° = 298°K (see table below) for these cases.

---

explicitly the "average heat capacity" of the system from T = 0°K to T°K. Unfortunately, I find that both S (which must have units of energy per mole per °K) and the term TS (which must have the units of energy per mole) are both frequently called "entropy." The inherent tendency to confuse TS and S followed by the totally different approach of Boltzmann (see below) has led to the attribution of almost mystical qualities to entropy and confusion with the second law of thermodynamics.

Salt	N x 3R (J/mole °K)	Cv at 298°K (J/mole °K)	Entropy (S at 298°K)[10]
NaF	49.9	46.8	51.3
NaCl	49.9	50.5	72.1
NaBr	49.9	51.4	86.8
MgF_2	74.8	61.6	57.2
$MgCl_2$	74.8	71.1	89.9
$MgBr_2$	74.8	70.0	117.2
AlF_3	99.7	75.1	66.5
$AlCl_3$	99.7	91.1	109.3
Na_2O	74.8	73.0	73
MgO	49.9	37.2	27.0
Al_2O_3	124.7	89.7	50.9

When we consider a chemical reaction

$$A \rightarrow B$$

the change in the entropy ($\Delta S(T)$) multiplied by the standard (ambient) temperature (T), accounts for the difference in energy associated with the different histories of the reactants (A) and products (B) as shown below:

[10] The differences between Cv avg and S, can be accounted for changes in the heat capacity and phase changes of the salts from 0 to 298°K.

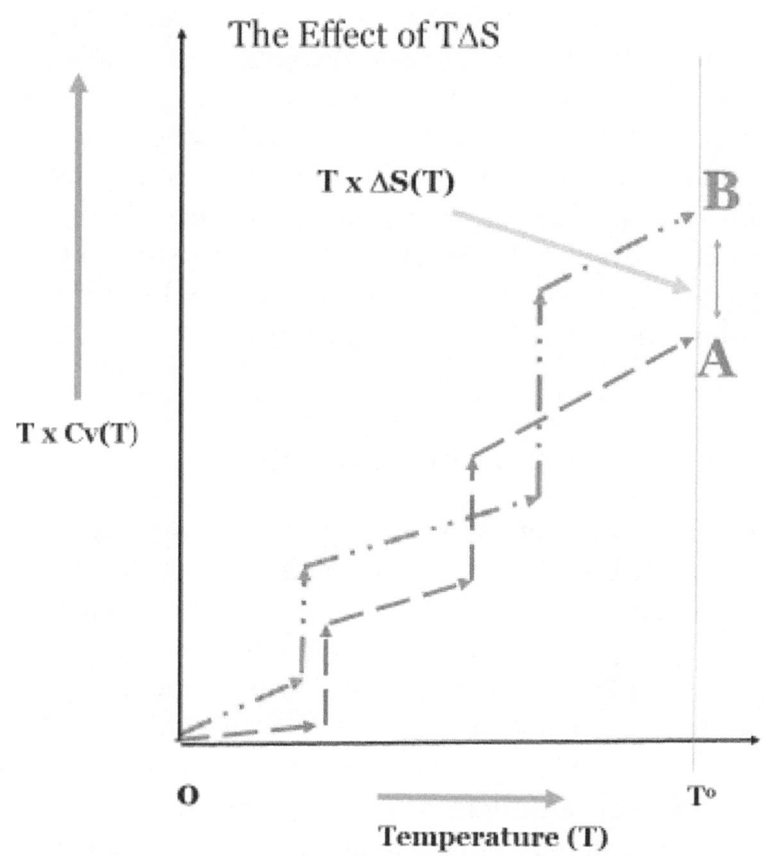

If you could transpose A into B without accounting for T x ΔS(T), you could build a perpetual motion machine.
© Parris 2015

If we did not have to account for the history of the reactants and products, it would be possible to create a perpetual motion machine.

5.0 Enthalpy

Enthalpy refers to the *relative energy* (kinetic, potential and work) stored in a system at a specific temperature and pressure. It may not be obvious, but the heat gained or lost by a system during a chemical or physical change at a constant temperature has no real relationship to the absolute energy in the initial system or the final system. The history of enthalpy reveals why experimentalists were fooled for a long time into thinking that the heat gained or lost by a system is indicative of its absolute energy.

5.1 The History of Calorimetry

The concepts of "heat" and "energy" had a long and convoluted history before 1900. For nearly 2000 years, "fire" was assumed to be a fundamental substance and in the late 1700s the *phlogiston theory* emerged in which heat was treated as a substance to be releases from combustible materials. This idea gave way to the *theory of caloric* in the early 1800s, which accepted heat as a form of energy but had no inference concerning its actual form as energy. As we saw above, the theory of molecular gases evolved during the mid-1800s and with it the idea that heat is kinetic energy of molecules.

Nonetheless, in 1761, Joseph Black (1728–1799) made an important theoretical observation: When ice and water are mixed, the temperature of the mixture always equilibrates at the freezing point (0°C). Moreover, when heat is added to a mixture of ice and water or a mixture of water and water vapor, the temperature does not change but the amounts of the two phases change. From these observations, he deduced that substances can store heat (which became the basis of the phlogiston theory). The heat stored in a chemical (without change in chemical composition) due to changes in temperature or change in phase at a constant temperature is called "latent heat." With his health failing, Black appears to have enlisted his academic colleague James Watt to build a calorimeter that depended on the melting of ice (measured by weight of water produced) to quantify heat. The ice calorimeter was designed with inner and outer chambers. The outer chamber contained a mixture of ice and water to maintain a constant temperature (0°C). The inner chamber contained only ice equilibrated to the temperature of the outer chamber (i.e., any water that formed in the inner chamber was allowed to drip out the bottom). Once the system was equilibrated so that melting of the inner ice was imperceptible, a vessel containing the experimental heat source (e.g., (i) a certain amount of water at an elevated temperature, (ii) a piece of heated metal at a certain temperature or (iii) a mouse in a bottle with oxygen) could be buried in the inner ice. After a period of time, the amount of water collected from the ice melt (indicative of the heat generated) could be compared to the corresponding change in the test sample (e.g., (i) the change in temperature of the water or metal in a sample or (ii) the amount of carbon dioxide produced by respiration or some chemical reaction). Controls would have to be run to account for any heating or cooling of the containers involved. Black gathered data until about 1766.

William Irvine (I743-1787) entered Glasgow University in 1756 as one of Black's students. He assisted Black in his latent heat experiments. He became the chemistry chair in 1770 and he is credited with expanding Black's idea of latent heat by introducing the idea of "heat capacity," i.e., the amount of heat required to change the temperature of a gram of a substance one degree centigrade.[11] Irving, unfortunately, introduced a confusion by assuming that the heat effects observed for chemical reactions (as well as physical changes) was entirely due to changes in the heat capacities of the reactants and products. However, this idea[12] presented a testable hypothesis that would trigger investigation in France when it arrived there in 1781.

John H. Magellan (1722-1790) was apparently instrumental in translating the British papers on heat into French. In the process, he summarized and organized materials from Black, Irvine, Adair Crawford (1748-1795) and Richard Kirwan (1733-1812) in his book *Observations sur la Physique* that appeared in two parts in 1780 and 1781; and he coined the term "specific heat" (*chaleur specifique*). The importance of Magellan's summary of Irvine's ideas was that they suggested that the heat of reactions could be predicted merely from determination of the specific heats of the reactants and Antoine Lavoisier's (1743-1794) understanding of combining volumes. Naturally, this would get the attention of Lavoisier in France.

In 1777, Lavoisier had begun overturning the phlogiston idea that solids are the source of phlogiston or that air can be dephlogisticated. Lavoisier and his assistant Pierre-Simon Laplace (1749-1827) were motivated to investigate the

---

[11] Not to be confused with his son also William Irvine (1776-1811) who included his father's work in essays written about 1805.

[12] Apparently, Irvine never attempted to test this hypothesis. Perhaps because you needed to know the combining ratios of reagents to make predictions.

possibility of a quantitative theory of phlogiston. They build an ice calorimeter similar to the one designed by Black in the spring and summer of 1782 and began measurements in the fall (when ice was available). After working through the cold months, Laplace read their co-publication *"Memoire sur la chaleur'"* in June 1783. In the background of this paper, Lavoisier summarized his caloric theory and Laplace summarized the heat/motion (*vis viva*) theory of Newton and Leibniz. Regardless of the theory, it is not surprising that their attempts to predict heats of reaction from specific heaths alone (Irving's hypothesis, see above) were disappointing. However, for simple changes of phase, (melting of ice to water) the theory was closer to correct (because the forced holding the water molecules together are relatively small).

Mainstream European chemists at the end of the 1700s were still left with the idea that heat was a material fluid called caloric. Nonetheless, imperfections in the theory of heat and thermodynamics were no impediment to the commercial success of the Boulton-Watt steam engine, which was revolutionizing industry. Thus, commercial success probably stabilized the intellectual *status quo*. Most authoritative voices who supported the motion (kinetic) theory of heat, did not object loudly to the idea that heat was a material fluid. The failure of caloric theory to provide a quantitative understanding of *heat of reaction* was not obvious because modern chemical theory about bonding and composition were in their infancy. It is interesting that there was almost no change in the theory of thermodynamics or heat from 1801-1819. Indeed, engineers still treat heat as a material fluid that flows from reservoir to reservoir.[13] Over

---

[13] The teaching of chemistry in colleges today is primarily in

the next 20 years (1820-1840), the kinetic theory of gases was finally accepted and this along with work on the heat capacity of solids and gases finally overwhelmed the caloric theory among chemists.

The use of the unit of heat "calorie" can be traced to the work of Nicholas Clément (1779-1841) who worked with Carnot on the mechanical equivalent of heat. In the period 1819-1824, Clément defined a Calorie as the *amount of heat needed to raise 1 kilogram of water 1 degree Celcius.* In modern usage, upper case "C" is used for these "kilocalories" although chemists usually use the upper case "C" and present data in calories (e.g, 246.8 calories). The term *Calorie* was not common in England until 1863.

In St. Petersburg, Russia, Germain Henri Hess (1802–1850) summarized the idea that the energy of a system is independent of the pathway needed to reach the system in 1840. He based his conclusion on an intuitive understanding of conservation of energy and experiments on the heats of solvation of sulfuric acid. The reaction of concentrated sulfuric acid with water is exothermic and Hess could use calorimetry (probably an ice calorimeter) to compare the heat generated at different degrees of dilution. From these experiments, he would quickly discover "Hess's Law"[14]:

---

support of engineering students. Thus, the chemistry textbooks retain a reverence for "heat reservoirs" and the Carnot cycle, which are of little interest to chemists and physicists.

[14] In 1918, Max Born (1882-1970) and Fritz Haber (1868-1934) applied Hess's law and developed the Born-Haber Cycle to determination the lattice energies of crystals that are experimentally inaccessible.

Acid + X Water → Dilute Acid + heat

Dilute Acid + X Water→ Very Dilute Acid + heat

Versus

Acid + 2X Water → Very Dilute Acid + 2 heat

Julius Thomsen (1826–1909) had followed Hess and gathered crude thermochemical data starting in about 1852.[15] Another 12 years passed before the (more precise) bomb calorimeter was invented about 1864 by Marcellin Berthelot (1827-1907). Berthelot was an organic chemist, not a theoretical physicist. And, he quickly conducted heat of combustion experiments that confirmed the work of Thomsen (+/- ~1%) and allowed him to draw conclusions about (i) relative stability of compounds and (ii) the heats of reaction. It was observed that most spontaneous reactions were also exothermal[16] (i.e., certainly combustion of organic compounds is exothermal). And, most chemists were led to believe that the heat of reaction was all that mattered in determining the direction of reaction. Throughout this period (1864-1885), Berthelot and Thomsen focused on heats of reaction determined by bomb

---

[15] George S. Parks. 1949. Some notes on the history of thermochemistry. *J. Chem. Edu.* :262-266.

[16] Berthelot invented the terms exothermal and endothermal.

calorimetry and ignored entropy (which had been discussed by Clausius in 1865, see above).

Tables of heats of reaction data could be applied to calculate the heats releases for reactions that had not been directly measured (i.e., application of Hess's law).

The late 1800s were a complex period in physical chemistry: For example, atoms were not fully established in the minds of physicists at the same time organic chemists were drawing three-dimensional structures of molecules; and although entropy had been described, heats of reaction were seen as the driving forces for reactions. Berthelot and others were trying to relate heats of reaction to chemical bonding (i.e., *affinity*)[17] The nature of chemical bonds was interpreted exclusively as an electrostatic phenomenon (Coulomb's Law).[18] Ironically, ionic salts were not realized to be dissociated in aqueous solution.

5.2 The History of Enthalpy

The term "enthalpy" is actually a rather recent invention. Apparently, Heike Kamerlingh Onnes (1853–1926) was the

[17] From the viewpoint of organic chemists, heats of reaction were mainly a tool to reach the goal of understanding chemical bonding (i.e., affinity).

[18] For example, chemist Hermann Kolbe (1818–1884) rejected structural chemistry because it is not explained by electrostatics. Rationalization of bond angles did not arrive until Linus Pauling (1901-1994) and his hybridization of atomic orbitals (1931).

person who actually coined the term (following the lead of Clausius) from a Greek root (ἐν θάλπειν) meaning "internal heat" and it was first used in 1909 by Alfred W. Porter.[19]

5.2.1 Defining Enthalpy from the Heat of Reaction

Although the history is somewhat murky, *internal heat* (symbolized by the letter H) had been considered by to Clausius in 1865 at the same time he defined entropy (see above):

> *"H denotes the heat actually present in the body, which, as I have proved, depends solely on the temperature of the body and not on the arrangement of its parts"*

Clausius, working with minimal understanding of chemical bonding, was obviously only considering the current kinetic energy of a system. Helmholtz seems to have had a broader interpretation of "internal energy", which included potential energy terms (like bond energy). Josiah Willard Gibbs (1839-1903) described a *heat function for constant pressure* (which he denoted by χ) in 1875:

$$\chi = \varepsilon + pv$$

A modern interpretation of this equation equates ε as the energy associated with chemical bonding and the second term is obviously the work done to create the system in the midst of a medium exerting a constant external pressure.

5.2.2 H vs ΔH

[19] Irmgard K. Howard. 2002. H Is for Enthalpy, Thanks to Heike Kamerlingh Onnes and Alfred W. Porter *J. Chem. Educ.* 79(6):697

We need to distinguish between enthalpy (H) and changes in enthalpy (ΔH). Interestingly, enthalpy (H) is actually not observable. We might define it at a standard temperature (T°) as

$$\mathbf{H(T^o) = E_0 + (T^o \times Cv(T^o)) + PV^o}$$

Where the external pressure is constant, but here is the fundamental problem:

Although we can measure the *heat capacity at a specific temperature* Cv(T°) in J/(mole x degree T), we know that it is not a constant (except for an ideal gas). Thus, while it may seem logical to multiply the heat capacity at a standard temperature (Cv(T°)) by the standard temperature (T°) to obtain an energy term (J/mole), this energy term actually results from an implied integration of the current heat capacity Cv(T°) over the temperature range (0 to T°) *assuming that it is a constant*:

$$\mathbf{T^o \times Cv(T^o) = \int_0^{To} (Cv(T^o)\, dT = Cv(T^o)\,[T^o - 0]}$$

Perhaps it is easier to see this error if I compare the common relationships of distance (D), velocity (v), acceleration (a) and time (t):
If I know the *average velocity* over a period of time I can calculate the distance travelled:

Distance = $V_{average}$ x time travelled
I can get the *average velocity* by integrating the instantaneous acceleration (velocity(t)/t) over the time of travel (0 to t) [analogous to what Clausius did to obtain the average heat capacity of a system (see above)].

But if I simply take the *current velocity* V(t°) and multiply it by the current time of travel (t°) the number I obtain has nothing to do with the *actual distance travelled*.

Similarly, if I take the current heat capacity (Cv(T°)) and multiply it by the current temperature (T°) it is tantamount to assuming that the heat capacity is constant throughout the temperature range (0-T°), which is not true *in general*.

On the other hand, if I take the difference in this number:

$$\Delta H(T^o) = \Delta E_0 + T^o \times \Delta Cv(T^o) + P\Delta V$$

I obtain a factor, ΔCv(T°), that is constant for the process under consideration and which can be merged with the change in bond energies to represent the *change in bond energy at the standard temperature* (ΔE_T):

$$\Delta E_T = \Delta E_0 + T^o \times \Delta Cv(T^o)$$

Thus, we can write (as so many have done before us):

$$\Delta H(T^o) = \Delta E_T + P\Delta V$$

For processes that occur at constant volume (e.g., in a bomb calorimeter):

$\Delta H(T^o) = \Delta E_T$

= the heat absorbed by the system (i.e., q)

Thus, while we cannot measure H experimentally, we can measure $\Delta H(T^o)$ by calorimetry.

The standard that was adopted was to assign the *internal energy of zero* to the common allotrope of each element at 298°K and 1 atmosphere of pressure. Using a series of reactions beginning with elements, e.g.,

$$C(s) + O_2(g) \rightarrow CO_2 + \Delta Hrxn$$

thus

$$\Delta Hrxn = [\Delta H_f^o \text{ of } CO_2] - [\Delta H_f^o \text{ of } C + \Delta H_f^o \text{ of } O_2]$$

$$\Delta Hrxn = [\Delta H_f^o \text{ of } CO_2] - [0]$$

$$\Delta Hrxn = \Delta H_f^o \text{ of } CO_2$$

Once this is done, the heats of reaction of simple compounds can be observed to construct an expanding network of heats of formation of all chemical compounds.

In practice, the approach which is used today is to *arbitrarily assign "zero"* as the enthalpy (i.e., standard heat of formation, $\Delta H_f{}^o$) for all elements in their stable allotropes at 1 atm of pressure and 298°K. Having made this decision, we are ignoring all the energy that went into an element between 0°K and 298°K. Hence, the "absolute enthalpy" of an element or compound must be calculated by subtracting the "entropy energy" (T^oS^o) from its relative enthalpy (ΔH_f^o) determined with the *arbitrarily assigned* values of enthalpy. Hence,

Absolute Enthalpy (G^o) = $\Delta H_f^o - T^oS^o$

5.2.3 The Relationship of Enthalpy to Entropy

An important point that should be made here is that the kinetic heat capacity of each major physical phase (e.g., solid, liquid or gas) is always increasing with temperature (such that $Cv(T^o)$ of a phase is larger than $Cv_{average}(T^{phase}$ to $T^o)$ of that phase).[20] Nonetheless, *when working with liquids or gases near ambient temperature (298°K),* the energy introduced in minor and major phase changes (between 0°K and T°) will *normally* make the average heat capacity $Cv_{average}(0$ to $T^o)$ larger than the current heat capacity $Cv(T^o)$:

Normally (liquids and gases):

$$Cv(T) < Cv_{average}(0\text{-}T)$$

As a result, the actual heat represented by the system (T^oS) will normally be greater than the heat calculated from the *current* heat capacity by multiplying by the *current* temperature $T^oCv(T^o)$. However, as we approach absolute zero (in the lowest energy physical phase) the current $Cv(T)$ approaches zero.[21] Nonetheless, it must be larger than the $Cv_{average}(0\text{-}T)$. When the crossover occurs

Crossover as T→0°K:

$$Cv(T^o) = Cv_{average}\ (0\text{-}T^o)$$

[20] See the data for carbon dioxide (above). However, note that changes in intermolecular bonds (e.g., hydrogen bonds) may constitute "minor phase changes."

[21] The motion of molecules and atoms is reduced (to the limit allowed by the Heisenberg uncertainty principle).

I will discuss the implications of this situation after I introduce free energy.[22]

6.0 Free Energy

6.1 Historical Development of the Concept of Free Energy

Josiah Willard Gibbs (1839–1903) was the first person to receive a PhD in engineering from a US university and his early work was on various practical applications in mechanical engineering (e.g., the design of brakes, gears and governors for engines). However, during a trip to Europe 1866-1869 he met Helmholtz; and after returning to the US (Yale University), he shifted his interest to mathematical physics and thermodynamics in 1871.

In 1873, he published a thermodynamic analysis of homogeneous materials using geometrical figures to capture the laws of thermodynamics. The publication in the *Transactions of the Connecticut Academy* was obscure, but he sent reprints to a number of people including James Clerk Maxwell (1831–1879) at Cambridge University. Maxwell constructed his own three-dimensional models based on Gibbs drawings and strongly approved the concept. Indeed, Maxwell was brilliant and may be the only person to ever actually understand Gibbs's presentation.[23] Unfortunately,

[22] I believe that this phenomenon may be associated with superconductivity (see below).

Maxwell was in poor health and soon died (1879); and Gibbs's ability to communicate his ideas did not improve.

His next paper (also published in the *Transactions of the Connecticut Academy* in two parts in 1875 and 1878) *On the Equilibrium of Heterogeneous Substances* received more notice (especially in Europe) and has been regarded as the founding of chemical thermodynamics. But it is so burdensome (700 equations) and complex that it had little practical impact until work by Helmholtz functioned as a Rosetta stone for chemists.

Indeed, the erroneous "Berthelot-Thomsen principle" (i.e., all spontaneous reactions are exothermal)[24] was generally accepted until at least 1882 when Helmholtz pointed out that it is only the "free energy" that determines the spontaneity of processes. Helmholtz was mainly working on mechanical systems in which chemical reactions (i.e., changes in chemical bonding) did not occur. In these systems, the pressure (e.g., in the cylinder of a steam engine) typically changes and his analyses do not assume constant pressure.

---

[23] I personally have tried to read Gibbs's papers and find them absolutely bewildering. Robert M. Hanson has published a much more readable paper "A unified graphical representation of chemical thermodynamics and equilibrium" (*J. Chem Ed.* 89:1526-1529 (2012). In this paper Hanson quotes Gibbs (lecture 11/27/1899) as saying "There is no great [practical] use of such a surface as this, but we are able to investigate some properties of thermodynamics by its aid." I heartily agree.

[24] This principle is actually true at "absolute zero" but needs to be corrected for energy involved in bringing the reactants and products to the ambient temperature of reaction. See Walther Nernst's (1864-1941) heat theorem 1906.

Chemists rarely approach problems this way. Chemists are usually much more interested in constant pressure conditions and most credit is given to Gibbs by assigning "G" to represent free energy. The resulting equation is simple and easy to use, but poorly understood:

$$\Delta G = \Delta H - T\Delta S$$

6.2 Interpretation of Free Energy at Ambient Temperatures

Most students, professors of chemistry and textbook authors never do more than memorize the Gibbs equation (above) and learn how to take data from standard tables, plug it into the equation and calculate the Free Energy (ΔG) of a reaction. The reaction will be spontaneous if the Change in Free Energy is negative, in equilibrium if it is zero and will go in the opposite direction if it is positive. I have spent many hours over the last 50 years trying to understand what Free Energy is and why this equation works. I hope I have it right now and I hope I can explain it to you.

Absolute energy must be conserved. Moreover, Clausius taught us how to calculate to *absolute thermal energy* (see above) of a system using the device of the average heat capacity over a temperature range (0 to T^o). Recall how the absolute thermal energy (Clausius's entropy) compares to enthalpy (H):

$$Cv_{average}(0 \text{ to } T^o) = \int_0^{To} (Cv(T)/T^o)\, dT$$

Versus

$$Cv(T^o) \int_0^{To} dT = To \times Cv(T^o)$$

If we assume that the average heat capacity $Cv_{avg}(0\text{-}T^o)$ will be larger than the current heat capacity $Cv(T^o)$, which is the most common case at ambient temperature (but see 6.3, below), we obtain the figure below. You will notice that the absolute energy (defined above) is the same as the Gibbs Free Energy (G) and that if it is negative (as it is in the Figure below) the system should be stable relative to isolated atoms. Moreover, it is clear from the figure why G = H-TS. If we compare two different systems, then $\Delta G = \Delta H - T\Delta S$.

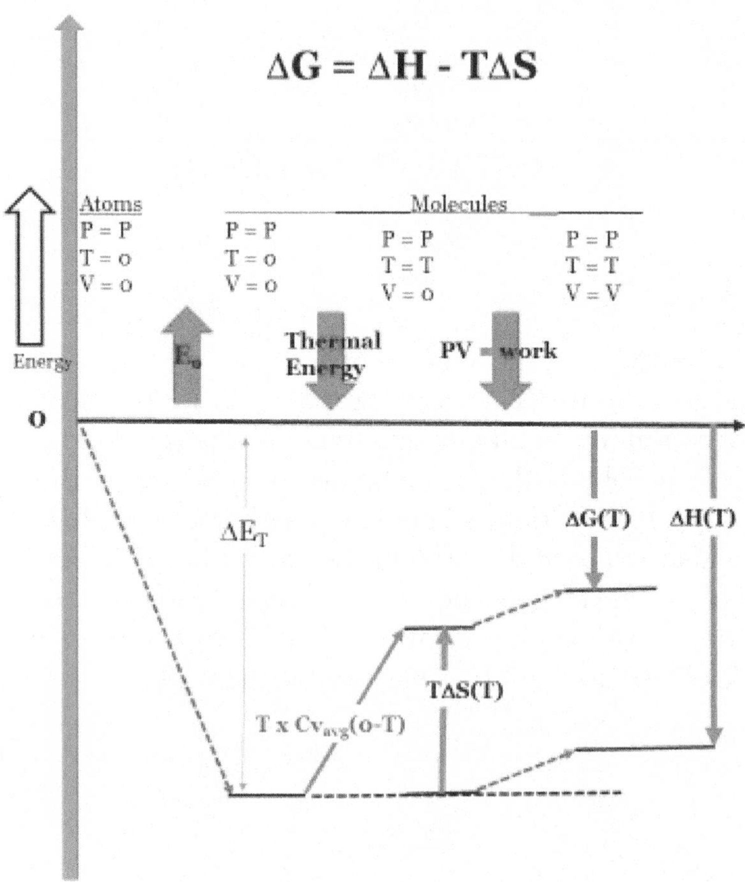

One way to interpret these results is that the experimentally observed heat (Q/n) we obtain from a reaction must be corrected for the fact that the two systems contained different amounts of heat to start with if we want to know how much energy is actually available to do physical, chemical or electrical work. In this view, enthalpy (H) of a system is divided into internal energy (U) and PV. Where U is only defined on a relative scale based as $\Delta U = \Delta H - \Delta(PV)$. For a (monatomic) idea gas, of course, there is no bond energy and all the internal energy is kinetic. But, for polyatomic molecules potential energy (i.e., bond energy and conformational potential energy of flexible molecules) and kinetic energy become inseparable.

Pedagogical Note

A simple way to present these ideas, it to explain that changes in enthalpy (ΔH_{RXN}) can be measured experimentally. But since we do not a priori know what the absolute energies of formation are we simply have assigned all elements in their most stable allotrope to have zero heat of formation at 298°K and 1 atm of pressure (i.e., $\Delta H_f^o = 0$ J/mole for pure elements). With this arbitrary assumption we can use experimental results and Hess's law to calculate relative standard heats of formation for compounds, ions in solution etc. While this system works well for calculating heats of reaction, if we want to determine that *actual relative energy of reaction,* we have to correct the enthalpies of formation for the (non-zero) energy (i.e., entropy energy) that was already there in the elements. Thus,

$$\mathbf{G^o = H^o - T^o \times Cpavg \text{ (in the range } 0\text{-}T^o)}$$

Or

$$\mathbf{\Delta G^o = \Delta H^o - T\Delta S^o}$$

I think this is brief, correct and easily understood explanation of the Gibbs equation.

To demonstrate the point that the TΔS term is a correction for the energy already in the system, you can look at tables of thermodynamic data for pure metals and salts that do not have a significant change in phase between 0°K and 298°K. In these cases, the standard entropy is usually higher than the theoretical heat capacity (n x 3R. where n = number of atoms in a mole) as discussed in section 3.8 above. The "hard metals" (Li, Mg, Al and the first-row transition metals except Zn) have standard entropies between 28 and 33 J/mole °K, which is close to the theoretical 25 J/mole °K. But in each family (column) the standard entropies increase as the atomic number increases, which may related to the electronic entropy (see section 8.4.2).

6.3 Interpretation of Free Energy Near Absolute Zero

At ambient temperature (e.g., T° = 298°K) many substances are liquids or gases. Even those materials that are solids at ambient temperature are often rendered into liquid form by dissolution in a solvent (which increases entropy) before conducting chemical reactions. Thus, (as discussed in 6.1) the *average Cv between 0°K and T°* is almost inevitably larger than the *heat capacity at ambient temperature.*

$$Cv_{average}(0 \text{ to } T^o) = \int_0^{To} (Cv(T)/T^o)\, dT$$

larger than

$$Cv(T^o) \int_0^{To} dT$$

Note that the heat capacity is the *slope* of the line in the figure. The energy calculated by multiplying each slope by T° will, of course, be larger for the larger slope. Thus, as shown above the difference is greater than zero:

$$Cv_{average}(0 \text{ to } T^o) - Cv(T^o) \geq 0$$

Such that the usual interpretation of the relationship of free energy, enthalpy and TΔS is observed (see graph above):

$$\Delta G = \Delta H - T\Delta S$$

But, as we approach absolute zero, we are likely to enter a physical phase where there are no *phase changes associated with the motions of atoms or molecules* in the range *0°K to T (i.e., as T → 0°K)*. Recall that the energy level separations are such that as the temperature of the material drops below about 30°K the molecules (or atoms) are going to be in their vibrational and rotational ground states (only retaining zero-point energy, Heisenberg's uncertainty principle) and in solids translational degrees of freedom are reduced to *librations* (long, low-energy vibrations) which themselves become very limited as the material approaches 3°K. So, as is well accepted, the Cv(T) approaches zero as T approaches 0°K. No more enthalpy can be lost from the system by rotations, vibrations and translations; and since the surroundings are (undoubtedly) warmer than the system, any exchange of photons (i.e., blackbody radiation) will tend to warm the system. Thus, the only way for the very cold material to decrease its free energy is for electrons to fall into lower energy states. This, is an exothermal *and* negative entropy event (e.g., decreasing the heat capacity). If the *average* heat capacity goes negative (i.e., the slope of the Cv/T line is more negative), we obtain the result shown below:

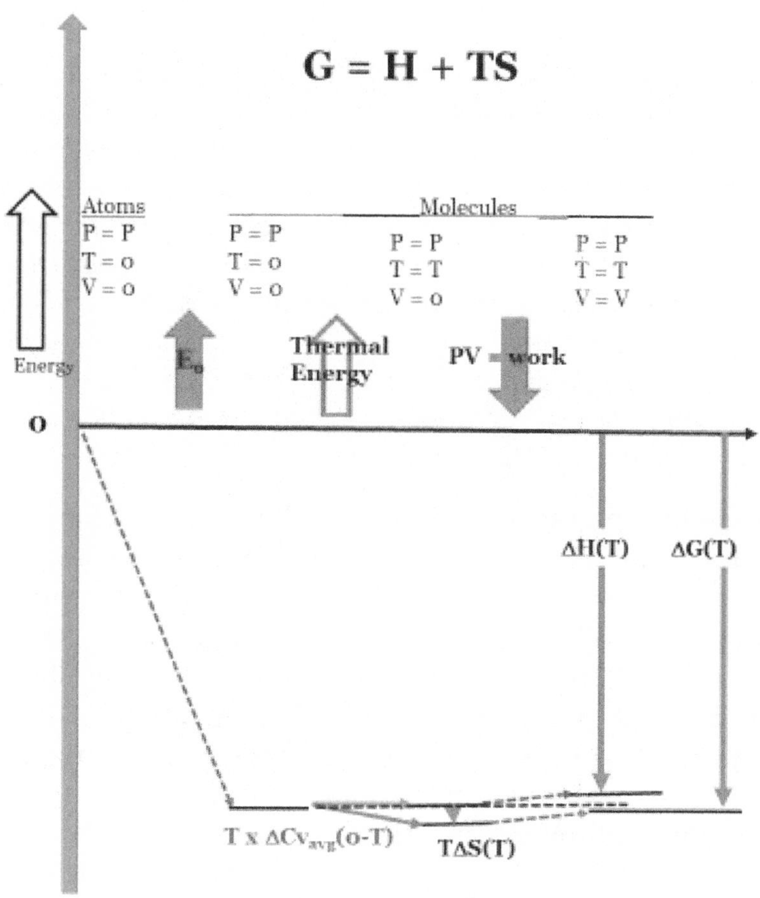

Copyright © 2017 George Parris

In 1849 [25], Gustav Kirchhoff (1824–1887) predicted the phenomenon of superconductivity. In his paper "On the Motion of Electrons in Conductors" (*Progg. Annal. Bd.*

[25] Before James Clerk Maxwell's (1831–1879) equations (1861).

102:529, 1857), he considered cases where the resistance of a conductor approaches zero or infinite[26]:

> *"In the previous paper* [1849] *I discussed the solution to equations (14) and (15) for the special case which is approached the smaller the resistance is made. I proved that in this case the electricity in the wire progresses like a wave in a taught string with the velocity of light in empty space. It is of interest to consider the opposite case which is approached the greater the resistance is made. I will do this here on the assumption that the two ends of the wire are connected with each other… In the first case the electricity propagates like a wave in a taught string, and in the second case it travels like heat."*

I could interpret this as follows: As the electrons of an electrical conductor (e.g., a metal) fall into lower energy states, the Debye wavelength increases until (in the ground state) the wavelength includes the entire physical dimensions of the conductor (i.e., the lowest energy state of a particle in a box, see section 8.4 below). In addition, the electrons presumably pair up.[27] A chemist might say that *the orbital is as large as the conductor and the electrons move freely in it,* resulting in superconductivity.

---

[26] Translation by P. Graneau and A.K.T. Assis. Kirchhoff on the Motion of Electricity in Conductors. *Apeiron*. 11:19-25 (1994).

[27] Solid state physicists talk about "Cooper pairs" of electrons and it is not clear to me how they relate to ordinary electron pairs familiar to chemists. It is not surprising that there should be some resonance in the vibration of the static matrix of cation cores (which would show an isotope effect) and the flow of electron pairs through that positively charged field.

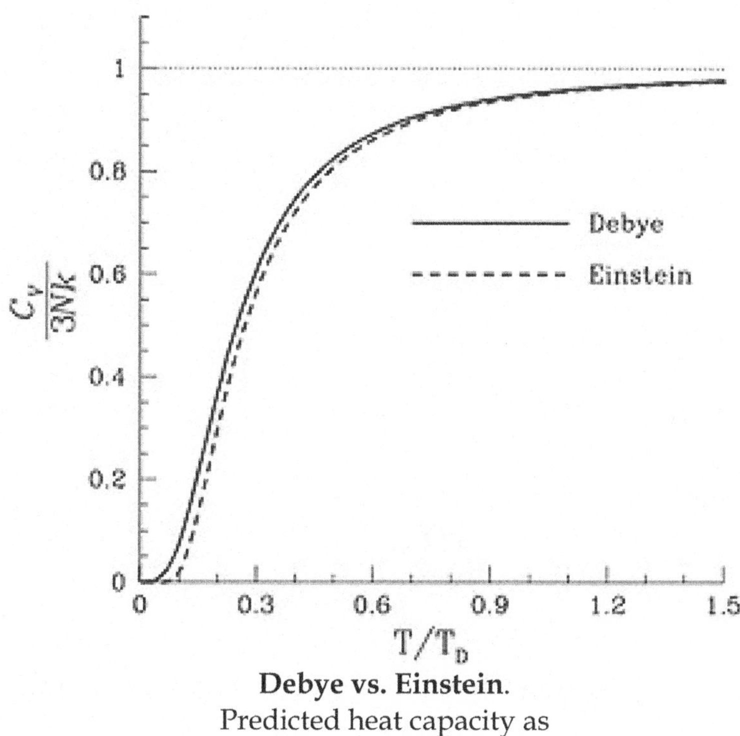

Debye vs. Einstein.
Predicted heat capacity as
a function of temperature.

Taken from English Wikipedia/Not eligible for Copyright

Regardless there is an abrupt increase in Cv at the critical temperature (Tc) for superconductivity, but the rate of decrease in Cv is more rapid below (Tc).

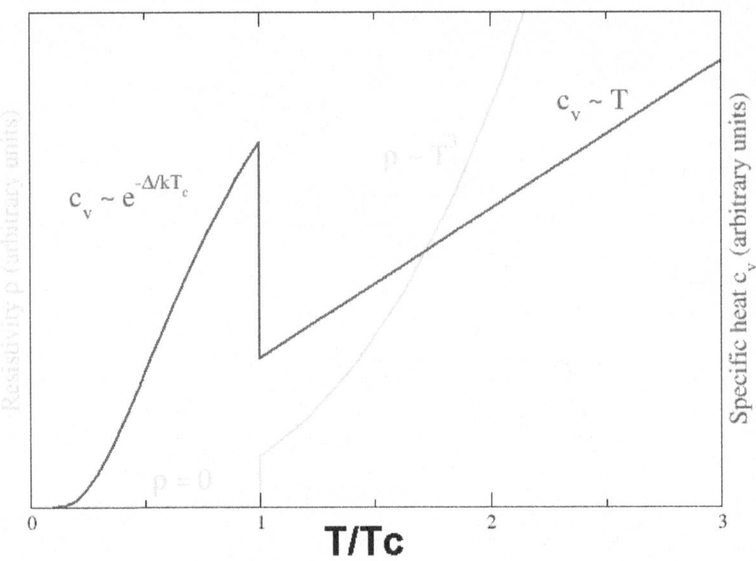

Drawing by Alison Chaiken, source Wikimedia Commons

7.0 Statistical Thermodynamics

7.1 Boltzmann

The above analysis shows the thermodynamic necessity of considering entropy as well as enthalpy in evaluation of the spontaneity of a chemical reaction. But it may seem odd that while the Free Energy (G) of a system is clearly a state function (i.e., it is independent of the path by which it was achieved), it is being defined by a specific pathway (series of events such as phase changes). Moreover, the elements (H and S) that make up the free energy should be independent of the pathway and *it should be possible to determine these functions without knowing the pathway*. This was the challenge that Ludwig Boltzmann (1844–1906[28]) set for himself.

Boltzmann was a theoretical physicist/ mathematician. He looked into the molecules and realized that the "history" of the transition of a system from T = 0°K to T = T° is "inscribed" in its energy levels. Recall, that the ability of a molecule to store energy is tied to the energy levels that are available to the molecule. Thus, Boltzmann reasoned, if he knew the energy levels available to a molecule, he did not need to know the pathway by which the molecule reached its current state.

7.1.1 Heat Flows from Hot to Cold

Between 1867 and 1872, Boltzmann focused most of his efforts on deriving Clausius's second law of thermodynamics from a statistical model. His model was embodied in what is called Boltzmann's transport equation (BTE), which considers what happens to the function describing a gas as it moves (flows, diffuses) and the molecules collide with one another. The issue came down to saying that if a gas were initially in an improbable state it would spontaneously move towards a more probable state; reaching the Maxwellian (i.e., most probable) state as the limit (i.e., at equilibrium). These arguments and resulting proofs were challenged with various objections, which Boltzmann addressed piecemeal as they arose (1872-1895). In the end, he conceded that the several lines of argument never completely proved his point and that ultimately it was a matter of intuition that a gas would proceed to the most probably state.

[28] Suicide.

> *"It can never be proved from the equations of motion alone, that the minimum function H [29] must always decrease. It can only be deduced, from the laws of probability, that if the initial state is not specially arranged for a certain purpose, but haphazard governs freely, the probability that H decreases is always greater than that it increases."* [30]

This is a probabilistic statement of the second law of thermodynamics.

7.1.2 Quantized Energy Levels

In section 3.3 above, we have already discussed the origin of the Maxwell-Boltzmann distribution function. It was first proposed by Maxwell in 1860 and adapted by Boltzmann between 1872 and 1877. Obviously, if the velocities of gas molecules are tied to the Gaussian distribution, so must their kinetic energies be tied to the Gaussian distribution; and by application of the equipartition principle, the energies in all degrees of freedom of molecules must be described by the Gaussian distribution. Since the events tractable by the

---

[29] Do not be confused, the "H-function" is not enthalpy as we have defined it; it refers to one of the several arguments that Boltzmann made between 1872 and 1895.

[30] Boltzmann, L. (1909), *Wissenschaftliche Abhandlungen*, Vol. III, p. 540, F. Hasenöhrl (ed.), Leipzig: Barth; reissued New York: Chelsea, 1969. As cited by Jos Uffink in *The Stanford Encyclopedia of Philosophy* is copyright © 2014 by The Metaphysics Research Lab.

Gaussian distribution are discrete, it is not surprising that Boltzmann began thinking along the lines of discrete energy levels associated with the degrees of freedom of molecules by 1877. In his 1877 paper entitled (roughly) "Concerning the relationship between the second main theorem of thermodynamics and the theory of probability" [31] he introduced the idea that the observable macro-state of a system (e.g., P, V, n, T of a gas) was determined by the summation of all existing micro-states (e.g., position and momentum) of all the molecules that made up the system. To enumerate the micro-states, he assumed that the energy of the molecules (e.g., monatomic gases with kinetic energy) was divided into discrete packages (energy levels) and that the system was described by molecules distributed statistically among the energy levels for each degree of freedom. The entropy of the system then was related to the probability of the sum of its micro-states and the system would tend to move to the most probable overall collection of micro-states.

The mathematical expression follows from this example:

Assume there are N identical gas molecules in a volume and each particle has **A** positions in the volume and **X** degrees of internal freedom. The total number of degrees of freedom will thus be **AX** to the power N:

$$\textbf{Total degrees of freedom} = (\mathbf{AX})^N$$

---

[31] L. Boltzmann: ¨*Uber die Beziehung zwischen dem Zweiten Hauptsatze der mechanischen W¨armetheorie und der Wahrscheinlichkeitsrechnung resp. den S¨atzen ¨uber das W¨armegleichgewicht, Sitzungsber. Kais. Akad. Wiss. Wien Math. Naturwiss. Classe* 76 (1877) 373–435.

If we define a state function (S) as a scaling factor (k) multiplied by the natural log of the number of degrees of freedom we have

$$\mathbf{S = k \ln (AX)^N = k\, N \ln AX}$$

This is simple enough, but how do we relate it to the macroscopic thermodynamic state functions (G, H, S) discussed above?

Consider what happens if the volume occupied by the gas is arbitrarily and instantaneously doubled (i.e., no energy is expended by the gas). But, the number of positions that can be occupied by the molecules has doubled (from **A** to **2A**).

The initial state function = $\mathbf{S_i}$

$$\mathbf{= k\, N \ln AX}$$

And

The final state function = $\mathbf{S_f}$

$$\mathbf{= k\, N \ln 2AX}$$

Thus

$$\mathbf{\Delta S = k\, N \ln 2}$$

How does this compare to the macroscopic state functions?

When the volume is instantaneously doubled, the temperature of the gas does not change, but as the volume V has doubled, the pressure of the gas will be reduced by a factor of 2. If this (S) is entropy, it should vary the same way that our heat definition of entropy varies.

We know that state functions should be independent of the path way, but we also know that moving along an isothermal path from V_iP_i to V_fP_f involves an energy change (ΔQ):

$$\Delta Q = \int_V^{2V} P\,dV = nRT \int_V^{2V} dV/V = nRT \ln 2$$

Thus, where n = N/6.02 x 10^{23}

$$nRT \ln 2 = \Delta S = k N \ln 2$$

Thus, if we define k = R/6.02 x 10^{23} we can write

$$\Delta Q = kT \ln 2$$

Or in general,

$$\Delta Q = kT \ln \Omega$$

But we know that

$$\Delta S = \Delta Q/T = \{kT \ln \Omega\}/T = k \ln \Omega$$

Where Ω is the number of micro-states.

7.2 Micro-States Make up the Observable Macro-State

In November 1900, Max Planck (1858-1947) used Boltzmann's idea of discrete energies to correctly predict the spectrum of blackbody radiation. And, he adopted the notation

$$\mathbf{S = k_B \ln W}$$

Where W is the thermodynamic probability of the macro-state. Following Boltzmann, W must be derived from the micro-states. In a simplistic model system (i.e., one degree of freedom and only a handful of molecules), **W** can be calculated for each micro-state as follows[32]:

$$\mathbf{W = N!/\Pi_i\ (n_i!)}$$

Where N is the total number of particles and ni is the number of particles in each micro-state. The entropy of the system is maximized (system is most probable) for the collection of micro-states when **W** is maximized. The constraint that is placed on each possible micro-state is that they each must account for exactly the energy in the system.

For example, consider a system of 7 particles (**X**), with a total energy of 13 units that each have one degree of freedom with equally spaced energy levels (0, 1, 2, 3, 4, etc.):

Energy Level	Micro-state 1
level 4	X 1 x 4 = 4
level 3	XXX 3 x 3 = 9

[32] A.H. Jungermann. Entropy and the shelf model: a quantum physical approach to a physical property. *J. Chem. Ed.* 83(11): 1686-1694 (November 2006).

level 2	
level 1	
level 0	XXX 3 x 0 = 0
Total energy	13
$S/k_B = \ln W$ $W = N!/\Pi(ni!)$	ln(5040/36) = **4.94** W = 7!/(1!)(3!)(3!)

Micro-state 2	Micro-state 3
X 1 x 4 = 4	X 1 x 4 = 4
X 1 x 3 = 3	X 1 x 3 = 3
X 1 x 2 = 2	XX 2 x 2 = 4
XXXX 4 x 1 = 4	XX 2 x 1 = 2
	X 1 x 0 = 0
13	13
ln(5040/24) = **5.35** W = 7!/(1!)(1!)(1!)(4!)	ln(5040/4) = **7.13** W= 7!/(1!)(1!)(2!)(2!)(1!)

Notice that the number of possible macro-states is limited by the constraint that each macro-state must have the same energy.

Of the three micro-states shown above (there are many other possibilities), micro-state #3 has the highest entropy. The table below, shows all the micro-states (identified as A-E) that have the same (maximum) entropy. These micro-states would be the most prominent micro-states in the system at equilibrium and they would be equally weighted because they all have the same entropy.

The "observable state" of the system shows the relative numbers of molecules in each energy state as predicted by the Boltzmann distribution (see section 3.3):

$$\mathbf{N_i/N_0 = g_i/g_0 \exp(-\Delta\varepsilon / k_B T)}$$

The observable system (i.e., macro-state) is the sum of all the micro-states:

	Micro-States					**Macro-State**
ε	A	B	C	D	E	**Observable State of System**[33]
7						
6					X	X
5		X	X			XX
4	X		X	X		XXX
3	X	X		XX	X	XXXXX
2	XX	XX	X	X	X	XXXXXXX
1	XX	X	XX	X	XX	XXXXXXXX
0	X	XX	XX	XX	XX	XXXXXXXXX
W	7.13	7.13	7.13	7.13	7.13	equal weight

7.3 Reconciliation of Boltzmann and Clausius

[33] Since we have only summed the most probable microstates, this is an approximation of the observable state of the system. But, the general pattern of a Boltzmann distribution is apparent. The less probable micro-states would be weighted proportionately less.

One of the interesting problems that Boltzmann's treatment of entropy presents is that it should be possible to deduce the entropy of a substance at any temperature without knowing anything about its phase transitions from absolute zero to the current temperature. However, Clausius's analysis of entropy requires a specific pathway of phase changes from absolute zero to ambient temperature (T^o). Of course, entropy is supposed to be a *state function* that is independent of the pathway. How can these ideas be reconciled?

The problem is actually in our fixation on physical states (e.g., solid, liquid, gas). We tend to view melting points and boiling points as fixed characteristics of materials (i.e., unrelated to thermodynamics). Moreover, because entropy (S and ΔS) is difficult to measure in most situations we have latched onto the relationship:

$$\Delta S_{\text{phase change}}$$

$$= \Delta H_{\text{phase change}} / T_{\text{phase change}}$$

in which the *temperature of the phase change* is implicitly treated as the independent variable and *entropy change of the phase change* is treaded as the dependent variable. This view is exactly backward. It is instructive to look at a plot of vapor pressure as a function of temperature. You will find that in spite of the dramatic onset of boiling, the vapor pressure curve is completely monotonic; the boiling point (T_{bp}) is an artifact of the external pressure. The melting point is a similar artifact, although less dramatic.

Instead of thinking of the phase transitions as fixed at certain temperatures, we need to realize that the different phases have different heat capacities and that *transitions between the phases occurs if (and only if) the ΔH for the phase change (e.g., solid to liquid) happens to exactly match the T(ΔCv(T)) of the two states of matter.* The entropy energy (TS) in one phase is accounted for as enthalpy (H) in the other phase:

$$\Delta G_{phase\ change} = 0$$

$$G_{phase1} = G_{phase2}$$

$$H_{phase1} - T(S_{phase1}) = H_{phase2} - T(S_{phase2})$$

$$T\ (S_{phase2}) - T(S_{phase1}) = H_{phase2} - H_{phase1}$$

$$T_{phase\ change}\ (S_{phase2} - S_{phase1}) = \Delta H_{phase\ change}$$

$$T_{phase\ change}\ (\Delta S_{phase\ change}) = \Delta H_{phase\ change}$$

$$T_{phase\ change} = \Delta H_{phase\ change} / \Delta S_{phase\ change}$$

In the following graphic, notice that the heat capacities of the two phases are different and the phase change occurs (only) when the difference between the two lines exactly matches the enthalpy of the phase change.

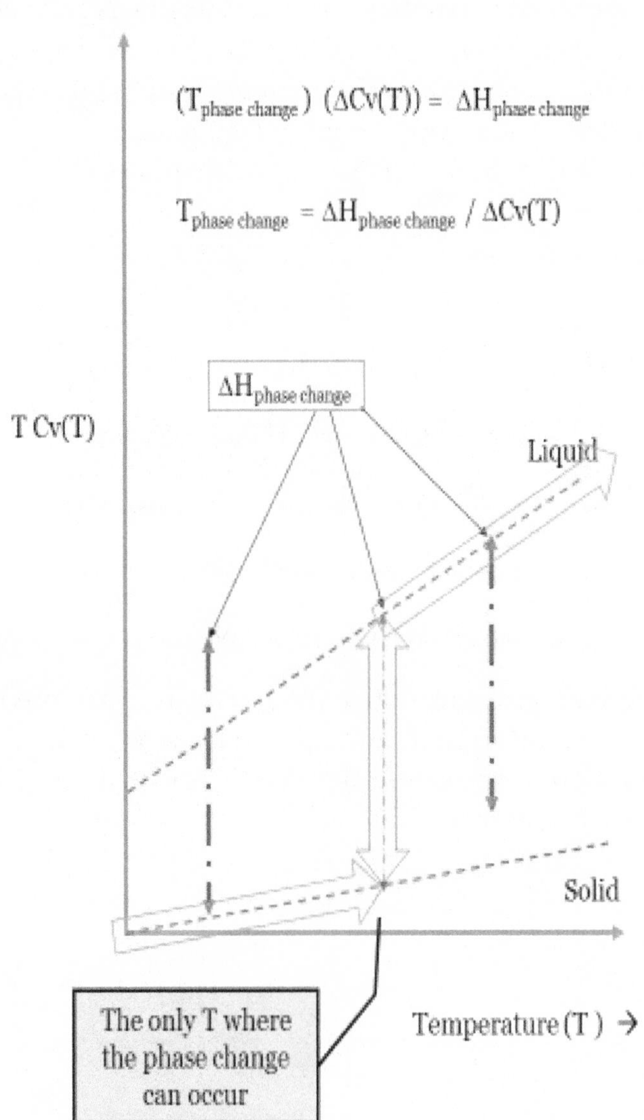
(T$_{phase\ change}$) (ΔCv(T)) = ΔH$_{phase\ change}$
T$_{phase\ change}$ = ΔH$_{phase\ change}$ / ΔCv(T)
ΔH$_{phase\ change}$
T Cv(T)
Liquid
Solid
The only T where the phase change can occur
Temperature (T) →

The transition between the phases (broad arrows) occurs if (and only if) the ΔH for the phase change happens to exactly match the T(ΔCp(T)) of the two states of matter. Energy must be instantaneously conserved!

Basically, the Clausius approach works, because the phase changes exactly compensate for the difference in Boltzmann's macro-states. Boltzmann's approach works because the current macro-state statistically integrates all the energy of all previous states (whatever they may have been).

8.0 Free Energy, Equilibria and Reaction Rates

8.1 Relationship of ΔG to the Equilibrium Constant (Keq)

The easiest way to see the relationship between the free energy change for a reaction and the equilibrium constant is to begin with the dynamic concept of the equilibrium. Consider the reaction

A ⇆ 2 B

(This particular stoichiometry has been chosen to allow demonstration of a point later on.) It can be made more explicit as follows:

A ⇆ [B:B] ⇆ 2 B

Here the symbol **[B:B]** represents the molecular assemblage of the two molecules of "B" brought into contact but without any modification of either molecule to accommodate formation of "A." In a gaseous reaction, the symbol represents a collision pair, in a liquid phase reaction the symbol represents two molecules inside a "solvent cage."

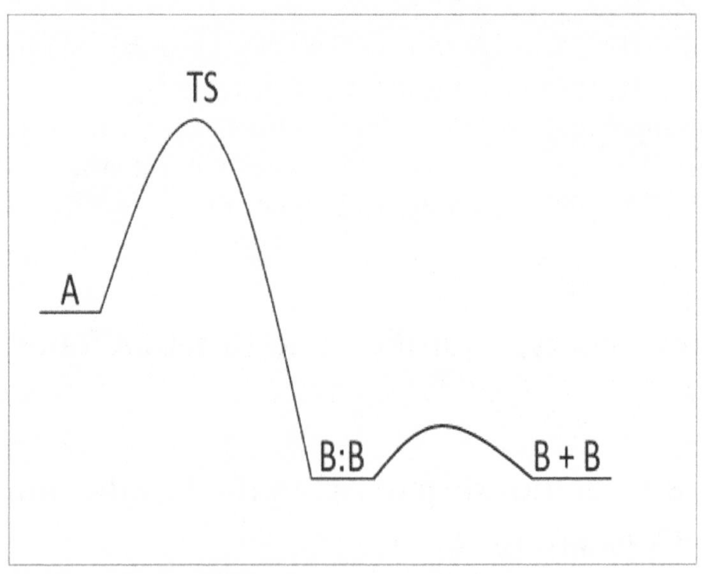

Adding another detail to the reaction path:

A ⇆ [Transition State ‡]⇆ [B·B] ⇆ 2 B

Boltzmann's equation relates the relative number (probability) for finding the assemblage of atoms in the various states (N_i/N_0):

$$[N_{A‡}/N_A] = e^{-(\Delta GA‡/RT)}$$

and

$$[N_{BB‡}/N_{BB}] = e^{-(\Delta GB‡/RT)}$$

"N" is the number of molecules in each quantum energy state.

At equilibrium, there must be the same number of transition states headed in each direction:

$$N_{A‡} = N_{BB‡}$$

Thus,

$$= e^{-(\Delta GA‡/RT)} / e^{-(\Delta GB‡/RT)}$$

$$= e^{(\Delta GB‡ - \Delta GA‡)/RT}$$

$$= e^{[(G‡ - GB:B) - (G‡ - GA)]/RT}$$

and recall that GB:B = GB

$$[N_{BB} / N_{A}] = e^{[GA - GB]/RT}$$

$$= e^{-[GB - GA]/RT}$$

$$[N_{BB} / N_{A}] = e^{-(\Delta G\ reaction)/RT}$$

Now, how is $[N_{BB} / N_{A}]$ related to the concentrations of A and B?

Although the absolute probabilities are unknown, the relative probabilities of finding BB or A in a volume of the reaction mixture have the same relationship:

$$[N_{BB} / N_A]$$

$$= [\text{Probability of BB/ Probability of A}]$$

and provided that all the concentrations are expressed in the same units (*e.g.*, moles/L, Molar), then the probability of finding two Bs in a certain volume of the reaction mixture is proportional to $[B][B] = [B]^2$ while the probability of finding A in the same volume is proportional to [A]. *Note that the probabilities are pure numbers (i.e., no dimensions of concentration)*. Thus,

$$\mathbf{[N_{BB} / N_A] = [B]^2/ [A]}$$

(dimension-less)

$$\mathbf{= \text{Equilibrium constant } (K_{eq})}$$

Thus,

$$\mathbf{[N_{BB} / N_A] = K_{eq}}$$

$$\mathbf{= [B]^2/ [A] = e^{-(\Delta G \text{ reaction})/RT}}$$

Solving for ΔG:

$$\mathbf{\ln K_{eq} = -(\Delta G \text{ reaction})/RT}$$

which rearranges to

$$\Delta G_{reaction} = -RT \ln K_{eq}$$

Where K_{eq} is dimension-less when [B] and [A] are in the same units.

8.2 Activity Coefficients

In the analysis above, It was assumed that the Free Energy of B:B and 2 B were the same. Suppose they are not.

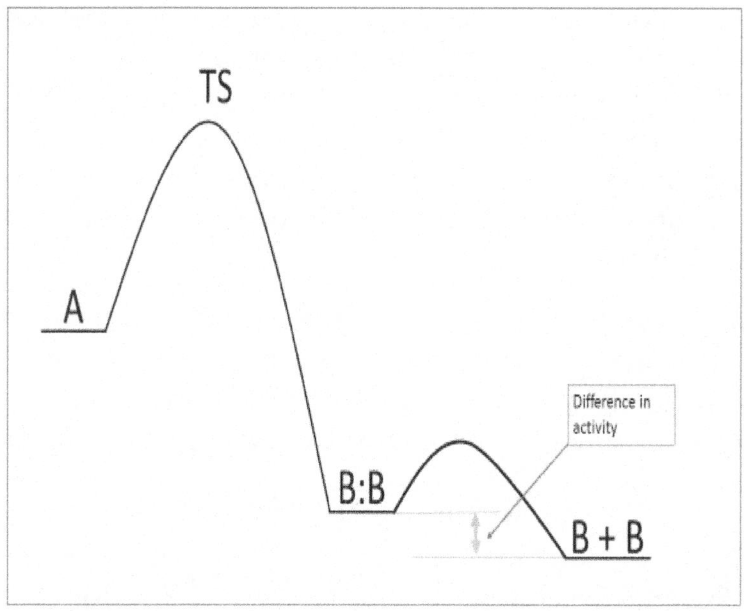

$$\mathbf{2\,B \leftrightarrows [BB]_{collision\ complex}}$$

If formation of this complex is not governed purely by chance, then the empirical equilibrium constant determined from the *analytical concentrations of reactants* and products will not yield the correct

$$\Delta G_{reaction} \text{ from } -RT \ln K_{eq\ (analytical)}.$$

The only reason that the collision complex would not form by chance would be that there are some interactions (i) between the reacting molecules and/or (ii) the reacting molecules and the solvent (or container) that changes the relative free energies. Note that every molecule in the system is subject to these sorts of variation when interactions with the solvent and other molecules is considered. Except when there are major changes in electrical charge or polarity (e.g., formation of ion-pairs), the effects of solvent tend to affects the reactants, transition state and products about the same way such that the relative differences (e.g., ΔGs) are not much affected. But, when ions or strongly polar molecules are involved, the analytical concentrations and the activities can be quite different.

We can correct the empirical analytical equilibrium constant ($K_{eq(analytical)}$) to the thermodynamic equilibrium constant ($K_{eq(thermodunamic)}$) by introducing correction factors called *"activity coefficients."* For example, using the equation above:

$$\mathbf{K_{eq(analytical)} = [B]\,[B]\,/\,[A]}$$

$$\mathbf{K_{eq(thermodynamic)} = \gamma_B[B]\,\gamma_B[B]\,/\,\gamma_A[A]}$$

We call the coefficients γ_i *activity coefficients* and the products γ_i [i] the *activities*:

$$\mathbf{a_i = activity\ i = \gamma_i\ [i]}$$

For certain types of interactions, we can calculate the activity coefficients. In particular, for ions, the activity is changed by the dielectric constant of the solvent and the ionic strength (concentration of other ions in the solution). The Debye-Huckle equation has been used to calculate these activities.

$$\ln(\gamma_i) = -\frac{z_i^2 q^2 \kappa}{8\pi\varepsilon_r\varepsilon_0 k_B T} = -\frac{z_i^2 q^3 N_A^{1/2}}{4\pi(\varepsilon_r\varepsilon_0 k_B T)^{3/2}}\sqrt{\frac{I}{2}} = -Az_i^2\sqrt{I}$$

Source:
https://en.wikipedia.org/wiki/Debye%E2%80%93H%C3%BCckel_equation

A is a constant that depends on temperature and solvent

$\mathbf{z_i}$ are the charges on the ions

I is the ionic strength

8.3 Free Energy of Reaction under Non–Standard Conditions

It must be recalled that the standard conditions assume chemical activities equal to their molar concentrations or pressures (atm); and solids and pure liquids have an activity of unity. Thus, if you are working in non-standard conditions you must correct for changes in the activities (see 8.2 above). This is done by correcting

$$\Delta Grxn = \Delta G^o rxn + RT \ln Q$$

Where Q is the reaction quotient (i.e., ratios of stoichiometric concentrations or pressures, products over reactants).

8.4 Practical Considerations

8.4.1 Experimental Determination of ΔH and ΔG

Making practical use of this information requires some ingenuity. The two pieces of information that are readily available from experiment are $\Delta H(T^o)$ and $\Delta S_{\text{phase change}}$ for (mp and bp). It is more difficult to obtain Cv for each phase between 0°K and T°. As noted above, it is a straightforward task to create a table of standard heats of formation ΔH_f^o (1 atm, 298°K) assuming that the standard enthalpy of formation of each element in its most stable allotrope is exactly zero under those standard conditions. From that, we can assign bond energies that are self-consistent. It is relevant that Coulomb's law can be used to independently calculate bond energies for ions (compounds and crystal lattices) that are in agreement with the bond energies determined thermodynamically (e.g., via the Born-Haber cycle).

The same approach is taken with free energies of formation ΔG_f^o. However, the standard free energies of formation are typically determined from equilibrium constants (see 8.1, above) not the heat of reaction. This is because as equilibrium constants approach 1, the error in measuring the analytical concentrations and calculating the equilibrium constant is less than the error in thermochemical measurements of enthalpy.

8.4.2 Experimental Determination of Entropy

It is relevant that tabulated values of entropy are absolute numbers, not relative numbers, and they are determined more or less as defined by Clausius. Modern techniques involve the use of differential scanning calorimetry to determine the heat capacities of various phases of materials from T° down to about 15°K. These components of the entropy (S°) can be combined with the contributions of entropy of phase changes ($\Delta H/T$) to obtain what is called the "non-Debye" contribution of entropy.

The entropy introduced between 0°K and ~15°K can be estimated by the method of Peter Debye (1884-1966). At low temperatures, the contribution of the lattice vibrations to the heat capacity of a solid is approximated by

$$Cp = aT^3$$

Where "a" is an empirical constant

This equation is consistent with the idea that the heat capacity and entropy are zero at absolute zero. The equation

$$(Cp/T)\,dT = (aT^3/T)\,dT = (aT^2)\,dT$$

is easily integrated from absolute zero to T_{low} (e.g., 15°K) to yield

$$\mathbf{S(0\text{-}T_{low}) = a(T_{low})^3/3}$$

It should be remembered that for metals (conduction solids) the electron heat capacity is larger than the lattice heat capacity at low temperatures. Thus, the Debye contribution to the entropy is usually negligible for **non-conducting** materials (insulators) at ambient temperature and it is dominated by electronic contributions[34] to entropy in **conducting** materials (metals) at low temperature (see 6.2 above).

9.0 The Arrhenius Equation: Prediction of Rate Constants

Most of the work to develop an equation to predict the rate constant for elementary reactions, i.e., the *Arrhenius equation* was done above. Start with:

A ⇆ [Transition State ‡]⇆ [B:B] ⇆ 2 B

The "*transition state*" is the highest point on the lowest energy pathway between reactants and products.

Typically, in the transition state, bonds are partially broken or partial formed, the bond lengths and angles are not in the optimum position for either the reactants or products. Once at the transition state (also called the "*activated complex*" especially when working with gas phase reaction), the chemical assemblage may either return to reactants or continue to products. Here, we are only going to concern ourselves with the forward reaction. The concentration of chemical assemblages in the transition state must be in

---

[34] The electronic heat capacity varies proportional to T at low temperature (see section 6.2).

equilibrium with the reactants (and products) at the reaction temperature.

The Boltzmann equation relates the relative number (probability) for finding the assemblage of atoms in the various states (N_i/N_0):

For the forward reaction (as written):

$$[N_{A‡}/N_A] = e^{-(\Delta GA‡/RT)}$$

We can re arrange this to

$$[N_{A‡}] = [N_A]\, e^{-(\Delta GA‡/RT)}$$

The rate of an elementary reaction of a transition state is determined by the rate of bond distortion, which is on the order of the *frequency of vibration* (assuming that activating collisions are not less frequent):

$$[N_{A‡}] \Rightarrow \textbf{products}$$

$$\textbf{Rate} = [N_{A‡}]\ \textbf{frequency of vibration}$$

Conventionally, the letter "*A‡*" for Arrhenius is used for the "*frequency factor*" since we do not *a priori* know what vibrations are involved or what their frequencies are.

Rate of formation of products

$$= [N_{A‡}]\ \textbf{frequency factor}$$

Rate of formation of products

$$= A^{‡}\, [N_{A‡}]$$

Of course, $A^{\ddagger}$ *is on the order of* 10^{12} *to* 10^{14} *per second*, typical of vibrations. We can now substitute for $[N_{A\ddagger}]$:

$$[N_{A\ddagger}] = [N_A]\, e^{-(\Delta GA\ddagger / RT)}$$

Rate of formation of products =

$$(A^{\ddagger}\; e^{-(\Delta GA\ddagger / RT)})[N_A]$$

Comparison of this equation to the ordinary rate equation

rate = (rate constant) [A]

we realize that the rate constant is given by:

$$\text{Rate Constant} = k_{rate} = A^{\ddagger}\, e^{-(\Delta GA\ddagger / RT)}$$

This is the Arrhenius equation.

We already noted that $A^{\ddagger}$ is related to the frequency of bond vibration. The exponential term, $e^{-(\Delta GA\ddagger / RT)}$, is the fraction of molecules that have the necessary energy to reach the transition state. One point that should be noted before leaving this topic is that this equation has the Free Energy of Activation ($\Delta GA^{\ddagger}$) rather than the "*energy of activation*" or the "*heat (enthalpy) of activation*." Free energy takes into consideration entropy factors (e.g., statistical considerations) such as the necessary orientations of the reactants in the activate complex.

We can replace ΔG with ΔH - TΔS and modify the equation as follows:

$$\mathbf{k_{rate}} = A^{\ddagger}\ \mathbf{e}^{-(\Delta GA\ddagger / RT)}$$

$$= A^{\ddagger}\ \mathbf{e}^{-(\Delta HA\ddagger / RT - \Delta S/RT)}$$

$$= A^{\ddagger}\ \mathbf{e}^{-(\Delta HA\ddagger / RT)}\ \mathbf{e}^{(\Delta S/RT)}$$

The entropy term can be considered an empirical "orientation factor" (ρ*) and the enthalpy term can be reformatted as the "energy of activation":

$$\mathbf{k_{rate}} = A^{\ddagger}\ \rho^{*}\ \mathbf{e}^{-(Ea / RT)}$$

In practice, the parameters $A^{\ddagger}\rho^{*}$ and **Ea** are determined experimentally by measuring the rate of reaction and calculating the rate constants at several temperatures. The equation is rearranged to solve for two unknowns:

$$\mathbf{\ln k_{rate} = \ln (A^{\ddagger}\rho^{*}) - Ea / RT}$$

If you have at least two values for the rate constant (k_1 and k_2) determined at two different temperatures (T_1 and T_2), you can solve the two equations for the two unknowns ($A^{\ddagger}\rho^{*}$ and **Ea**). Deviation of the first term from ~10^{-13} per second is usually rationalized qualitatively by invoking entropy (orientation) factors. In the case of gases at low pressure or bimolecular reactions in very dilute solution, the frequency of collision may be slower than the frequency of bond vibration and will determine the rate of reaction.

Made in the USA
Coppell, TX
14 February 2022

73583081R00062